D1367384

Florida, My Eden

Florida, My Eden

by
Frederic B. Stresau
F.A.S.L.A.

FLORIDA CLASSICS LIBRARY
PORT SALERNO, FLORIDA 34992-1657

For information:
Florida Classics Library.
P.O. Box 1657, Port Salerno, Florida 34992

Library of Congress Cataloging in Publication Data.
Stresau, Fredric B., 1986 Florida, My Eden. Exotic and Native Plants for Use in Tropic and Sub-Tropic Landscape. First Edition, includes index.
Catalog Number: 86-090532
ISBN 0-912451-19-X (Softcover)

COVER DESIGN AND PAINTING BY JAMES N. BAKER

PREFACE

Florida has been good to me, it has given me great opportunities and boundless rewards in my chosen profession. Even though it sounds like the title of a novel "Florida, My Eden" had to be the title of this book for Florida has been, and is truly my Eden.

In reflecting on the past, I attribute a modicum of my success by recognizing the geography, geology, climate and the existing ecology of the client's property. I also worked with a complete understanding of the client's wishes, the scope and the uses of the land.

The development of successful composition of plant materials seems to me, to be likened to the dressing-out of the naked figure, whether formal or informal and whether it is a natural selection of garb as opposed to a dress of state importance.

DEDICATED TO MY LOVED ONES

INTRODUCTION

Florida's geology, when first viewed, leaves something to be desired, with the southern half of the state varying from sea level to not much more than a plus 30 feet at the high. Yet, this "flattish" land contains a good number of flora and fauna in microclimates that make up individual environments. These are called communities and have special names describing them. Listing source of these communitites in major categories we have:

1. Coastal Dunes
2. Hammocks
3. Fresh water swamps or wetlands
4. Salt water wetlands
5. Sand pine scrub
6. Dry pine land
7. Pond and river margins

Each of the above has its own list of plants and fauna that rarely leaves its own community.

From Sarasota, Florida on the Gulf of Mexico draw a line, a short distance inland, South to Naples, thence to the ten thousand islands, around the Keys, then Northerly again close along the Atlantic Coast to approximately Vero Beach. This is the State's warmest temperature zone.

By virtue of the warm waters of the Gulf Stream and balmy trade winds, we enjoy a climate in this area very close to that of Cuba, Haiti and Puerto Rico. There is also an area on the South and East side of Lake Okeechobee which is traditionally in this warm belt. Rich muck land deposits that are part of the lake basin and winter winds tempered by the warm lake waters create an extended growing time to make it the "salad bowl" of Florida.

Over sixty-one percent of our plant species in this region are tropically oriented. The balance of the native species have drifted down from the temperate zones of the Southeastern

United States and Upper Florida. And too, since pioneer days, earlier settlers, seafarers, world travelers and later Federal plant introduction stations have been responsible for importing the great bulk of exotic plants which dominate today's nursery sales.

The movement to use native plants was rather obscure for years until the Federal Agencies forced awareness of water and energy conservation in the 70's and 80's. This was compunded by new concerns over the shrinking size and number of endemic plant and animal ecologies within our cities, their peripheral development areas, and rural sectors. The ability of true native plants, once established, to survive without special care and watering has propelled a new type of nursery into existance and another look at landscape design.

In this book, there is an effort to present the plants best suited for good landscape design. Every subject was weighted and evaluated on its present or future part of the whole in producing a happy and sucessful design. Not included are the "trash" trees, such as Australian Pines, Brazilian Peppers and Melaleuca Trees, sometimes called "Punk" Trees. As well, there are trees which are messy and dirty or ones with weak root systems or subject to wind damage.

Included are natives, which at present may be difficult to obtain, but given three to five years, will be available in several sizes. Some of them are more than qualified to present a superior picture over some of our imported exotics, and many are cold hardy and maintenance free.

There are a few new introductions that have been adjudged to fill a void in many categories, having proved themselves in length and color of bloom, controllable sizes for ease of maintenance, resistance to disease, tolerance of deep shade or cold resistance. From out of the woodwork, a few old timers have been resurrected to re-establish some knowledge of those that for a number of reasons have disappeared from todays nurseries, such as Blue Sage, Dwarf Malphighia, the single flowered Pin Wheel Flower or Crepe Jasmine, even plants of the Soft-tipped Yucca which are no longer popular.

ACKNOWLEDGEMENT

Many thanks to all my friends and colleagues who have given me their advice and help. John D. Longhill, landscape architect, who worked jointly with me on the color photography for over a year and who contributed so much to the quality of the work. Dr. Derek Burch, University of South Florida, now a Botanical and Horticultural consultant, has aided the author in bringing modern and current nomenlature to this book. Any flaws in the collating of Latin and common names are the writer's errata, not Dr. Burch's.

Dr. Roy Woodbury, Professor Emeritus, University of South Florida, Botany, Ecology and Taxonomy, mentor and lecturer on flora of the southern region of Florida. Dr. Joseph T. Bridges, Florida Institute of Technology, Plant Ecology and Botany, who aided in the nominal derivation of the Latin and Greek names. Nancy Angle, University of Florida, Ornamental Horticulture, for her ideas and inspiration in the field of color in the landscape of South Florida and a great supporter of this book.

Maggy Hurchalla, a leader and knowledgeable instructor to the lay naturalists on so many field trips of the natives on the Treasure Coast and elsewhere. Steven Adams, Loxahatchee Nursery, Stuart. Edward Hayslip, Sun & Shade Nursery, Fort Pierce. Hugh F. Forthman, Native Tree Nursery, Homestead.

To Lucy Mott Hupfel for her donating time in outlining the methods and pitfalls of book production and publication. My gratitude to Mrs. Trude Farrell, my firm's secretary, who always found time in her busy schedule to work on typing and retyping, of coordinating sections, coding lists, etc., etc. And finally to Mrs. Lyse Middleton for her industrious and faultless computer typing of my chicken scratching manuscripts.

ABOUT THE BOOK

FLORIDA, MY EDEN was written as a reference book on tropical and subtropical trees, shrubs, vines, ground covers, and accents. Mr. Stresau has also touched on cold tolerant native plants due to our recent climatic cold cycle. Each of these categories was selected by the author for its beauty and adaptability to our Florida climates.

Mr. Stresau has covered the genus, specie, Latin and common names as well as size, form, texture, leaf, flower, fruit, geographic location, culture and use of each piece of flora illustrated in this book.

ABOUT THE AUTHOR

Frederic B. Stresau has gained national recognition as a highly creative Landscape Architect. A graduate from the University of Illinois, certified to practice in New York, Michigan and Florida. In June, 1964, Mr. Stresau was elevated to Fellow in the American Society of Landscape Architects. He was selected by the Governor to serve on the first committee to establish the licensure for the State of Florida. He has accumulated, over a long and successful career, awards by the dozens.

Included are the American Association of Nurserymen, Inc., National Award of Merit, Americana Hotel (1967) San Juan, Puerto Rico; The National Award of Honor (1979) The Miami Airport, by the American Society of Landscape Architects; and more recently in 1985, The Florida Nurserymen and Growers Association Award of Excellence for the clubhouse and grounds at the Moorings, Vero Beach, Florida.

Mr. Stresau founded the Landscape Architectural Firm of Stresau Smith and Stresau, P.A., and served as President and Chairman of the Board until he retired. Upon his retirement, he began working as a Landscape Consultant. Mr. Stresau resides with his wife in Stuart, Florida.

NOMENCLATURE

The first name in Latin is the genus name and the initial letter is always capitalized. The second name is the species name and is usually lower case, rarely capitaized as per proper name. Clonal or English varietal names have an initial capital letter only and are embraced in single quotations.

Common names are set in capitals. Under derivations the botanical name is set in italics followed by the meaning which is given in book type.

Basically the International Code for the Nomenclature of Cultivated Plants now takes precedence over the older Standardized Plant Names, American Joint Committee on Horticultural Nomenclature 1942. Many name changes have occurred and they probably will continue to change over the years to come.

Dr. Derek Burch has been the advisor on much of the Horticultural and Botanical answers for nomenclature in this book.

CONTENTS

PALMS

The selection of Palm subjects discussed in this section no longer list the great number of species we might have written about twenty years ago. Today the less interesting Palms, the really tender ones decimated by the series of recurrent freezes and cold snaps, many of the ones subject to lethal yellows and other maladies have been ruled out. It was once brought to my attention by a client, a lady land developer in Ft. Lauderdale, who positively exclaimed, "Don't you ever forget that you are in the winter playground of America and arriving tourists shed their hats, furs and overcoats, get out the golf clubs and play under a canopy of waving palms. That makes Florida!"

Our State tree has been designated as Sabal Palmetto which we choose to call the Sabal Palm. It's an all-Florida native that now is used by the truckload (up to 40 trees per trailer). There are concerns about its over-use. But, like the Slash Pine, only a few years sees it replenish itself in great quantities as long as we still have open land.

The Palm is not a tree in the regular sense in that it has no dead wood forming the rigid trunk of broad-leafed trees, nor does it have layers of bark which cover the growing layer of tissue called the cambium layer. Instead, one might liken the trunk of a Palm to a thousand small soda straws which pass the sap to the growing bud or heart and attendant crown at the top. Thus, the entire trunk can be considered alive and injections of spikes of tree-climbers and nails result in creating cavities.

Acoelorraphe wrightii

PAUROTIS

acoelorraphe: minute valve in
floral parts

wrightii: Named for C. Wright
American Botanist

Size:	To 50 feet
Form:	Soda-straw cluster, clutched at the ground line, topped with compact crowns 5 feet in diameter of fan leaves for each stem.
Texture:	Fine
Leaf:	Orbicular, 2-3 feet across. Divided below the middle into narrow segments, green above & silvery below.
Flower:	Broom-like cluster
Fruit:	Red, ⅓ inch long, globose. Stem or trunk is plus or minus 4 inches in diameter and matted with a woven red fiber similar to burlap.
Geographic Location:	S.W. Everglades hardy to 20°F. W. Indies and Bahamas, Central and South Florida.
Dormant:	Evergreen
Culture:	Requires more than average moisture. Thrives on marl land, but will grow well on sandy soil. Fertilize in Spring and Summer. Medium growth. Treat with basic copper as a drench for budrot and fungus problems of the bud.
Use:	Our most handsome multistem palm is used at building entrances, on roof decks and as individual specimens in the landscape. Night scape lighting is outstanding for this palm.

Arecastrum romanzoffianum

QUEEN PALM

Arecastrum: Areca-like

romanzoffianum: For N.
   Romanzoff of Russia

Size:	To 40 feet
Form:	Vertical clear gray trunk plus or minus 12 inches in diameter. Topped by arching pinnate fronds 12 to 15 feet long in a symmetrical crown to 20 feet diameter.
Texture:	Medium
Leaf:	Fronds are plume-like arching with leaflets hanging from the mid-rib like a fringe.
Flower:	6 foot long spray issuing from a long spathe.
	1 inch diameter orange color, in quantity, globe-shaped.
Geographic Location:	South America to Orlando, but tender in North Florida.
Dormant:	Evergreen
Culture:	Tolerant of sandy soils, full sun, good drainage is *extra* important, fast growing. Look for poor crowns showing up in some nurseries. Use and specify, heavy, robust crowns. Pests: Scales, bud rot, mineral lack in calcareous soils.
Use:	For a natural arrangement select 3 trees of staggered sizes planted in a triangle about 10 feet apart. One of Central Florida's favorite street trees spaced 35 to 40 feet on centers. Crown rot can be controlled by direct applications of copper sulfate in the bud on two week intervals.

Caryota mitis

TUFTED FISHTAIL PALM

Caryota: Greek name first
    applied to Date Palm

mitis: Mild

Size:	25 feet
Form:	Clump growing by virtue of many suckers, upright in form.
Texture:	Medium
Leaf:	Leaf blades segmented into triangles resembling fancy gold fish.
Flower:	Inconspicuous in scurfy much-branched spadices.
Fruit:	Globose drupes ½ inch in diameter, bluish-black in color. Upon flower and fruit emergence the life of the tree begins its slow decline and death.
Geographic Location:	Burma to Malaya. Tropical, South Florida only.
Dormant:	Evergreen
Culture:	Well adjusted to thrive in deep shade or full sun. Prefers rich loam. Needs ample moisture. Rapid growth. 20 years is an average lifetime.
Use:	C. mitis has little leaf drop and thus is a good subject around swimming pools for screen use. As a patio specimen feature it as an individual and use it with illumination. Hard to beat for its tropical atmosphere in an evening garden area. Also a popular tubbed subject on verandas, porches and in hotel lobbies. Limited for interiors because of the pests red spider and mites.

Chamaedorea erumpens

BAMBOO PALM

Chamaedorea: Greek for dwarf and gift.

erumpens: for breaking through.

Size:	10 feet by 3 feet.
Form:	6 or more feet bamboo-like canes with lacy pinnate fronds in symmetrical loose crowns.
Texture:	Fine
Leaf:	Pinnate fronds 15 to 20 inches long. Canes bamboo-like ½ inch in diameter, drooping leaves.
Flower:	Cluster of 6 wirey stems plus or minus 6 inches long producing some 50 miniature flowers, yellow.
Fruit:	Uncommon
Geographic Location:	Honduras. South Florida only, Tropical.
Dormant:	Evergreen
Culture:	Rich fibrous soil, well drained, moist, shady. Needs an overstory or canopy as a cover.
Use:	This palm is usually relegated to interior tubs or planters. As an outdoor subject it is in its environment around pools, ponds and water features such as fountains and rivulets, but in the shade and protected from winds.
Other:	C. seyfritzii — more robust, stands full sun.

Chamaerops humilis

EUROPEAN FAN PALM

Chamaerops: Greek for dwarf bush

humilis: dwarf

PALMS

Size:	15 to 20 feet by 15 feet, some forms remain dwarf to 5-6 feet extremely variable.
Form:	Cluster forming clumps developed from offshoots, curving as they rise, crowned by fan-leaves.
Texture:	Medium
Leaf:	Green to bluish-green, palmate on spiny stalks. Fan leaf is pleated, leaf 2-3 feet across, stiff.
Flower:	Flowers are hidden from view
Fruit:	Spadix occurs among the leaves, ovoid fruit ½ to 1½ inches long, brown or yellow, somewhat succulent.
Geographic Location:	Mediterranean. Entire state of Florida. Reported to have survived 6°F in California.
Dormant:	Evergreen
Culture:	Drought and wind resistant. Feed and water in growing season to speed up growth. Pest: An as yet unidentified disease kills in crowded nursery conditions.
Use:	As a tubbed specimen. As an accent plant at entrances and in planter boxes. As an armed barrier screen. An 8 to 12 foot specimen is a feature plant as sculpture.
Other:	The species is variable, some are named.

Chrysalidocarpus lutescens

ARECA, BAMBOO PALM

Chrysalidocarpus: Greek for golden fruit

lutescens: Becoming yellow

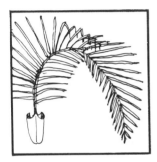

Size:	30 feet
Form:	Many canes gracefully issuing from a central ball silhouetting feathery fronds about the crown.
Texture:	Fine
Leaf:	Arching pinnate fronds forming symmetrical tops for each of the yellow-green 6 inch diameter canes.
Flower:	Inconspicuous brooms hidden in the foliage.
Fruit:	Oblong ¾ inch in diameter, dark brown.
Geographic Location:	Madagascar, South Florida, Tropical. At 28°F 90% of foliage brown and dead. Recovery takes 2 years.
Dormant:	Evergreen
Culture:	Rich sandy loam, well drained. The juvenile palm prefers shady conditions until 7-10 feet. Maintain good moisture in the dry season. Pests: scales and sooty mold.
Use:	Its fullness makes it a popular pot subject for outdoor paved areas as well as for interiors in shopping malls, hotel lobbies and commercial buildings. It is valued as a specimen and also when planted as screening or mass planting. A handsome background in mixed informal plantings.

7

Cocos nucifera

COCONUT

Cocos: Portugese for monkey

nucifera: nut bearing

Size:	40 to 60 feet
Form:	25 fronds each 12 foot long crowning a full fibrous stem.
Texture:	Coarse to medium
Leaf:	Pinnate fronds dark to yellow-green tapered to pointed tip, recurving.
Flower:	Broom-like issuing from a low on the crown, flowers occur on the panicle-stems, straw-colored.
Fruit:	The coconut. In the Malayan golden, the fruit color is a handsome muted orange. In the yellow Malay the color is lemon.
Geographic Location:	Malaya. South Florida. Tropic Asia and many Pacific Islands.
Dormant:	Evergreen
Culture:	Sandy loam. Moderate drainage — full sun, moisture. Fertilizing stimulates rapid growth, use in the spring only.
Use:	Since the advent of the disease "Lethal Yellows" Florida is using the resistant Malayan strain which is identified by the colors yellow, green and golden. Occasionally called Pygmy Coconut. Other than being characterized by a straight trunk, this strain is very close to matching the Jamaica tall. Of the 3 colors the "Green" is the most rugged and attractive foliage wise. The variety Maypan: a hybrid of the Malayan times the Panama Tall is the most robust and rapid growth, all are salt tolerant.

Cycas circinalis

QUEEN SAGO

Cycas: Greek name for a Palm
    Tree

circinalis: Coiled

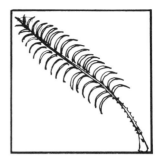

Size:	20 feet, usually half this size.
Form:	Stout, shortish trunk 6 inches in diameter with a crown of pinnate curving fronds 9 feet long by 1½ feet wide, tapering, shiny green and black.
Texture:	Fine
Leaf:	See above
Flower:	Female, velvety, modified leaves 1 foot long within foliage. Male, flower is a brown cone up to 2 feet by 5 inches wide.
Fruit:	Orange colored, 6-12 seeds per plant the size of bantam eggs.
Geographic Location:	Asia and South Pacific. South Florida, Tropical.
Dormant:	Evergreen
Culture:	Impartial to most soil types. Adjusts to sun or shade, prefers good drainage. Average to little care once established. Slow growth. Subject to leaf scale.
Use:	Most often used as streetside door-yard specimens. Also used in large tubs. Not fitted into landscape designs very successfully in the past, and not available in quantity.
Other:	Cycas revoluta; a 3-4 foot spread, relative, can resist 28°F temperatures.

Howea forsteriana

KENTIA PALM

Howea: For Lord Howe in
South Pacific Island

forsteriana: For a Mr. Forster

Size:	35 feet, crown spread to 15 feet
Form:	Single stout trunk with pinnate leaves.
Texture:	Medium
Leaf:	To 10 feet long. Leaflets extend horizontally till near the tip where the leaflets turn down.
Flower:	2 or 3 parted flower spike — less than 2 feet long.
Fruit:	Narrow
Geographic Location:	Lord Howe Island, South Pacific. South Florida, Tropical
Dormant:	Evergreen
Culture:	Soil should be well prepared with rotted stable or cow manure in a well drained location. Partial shade with protection from winds. Discourage grass from growing on roots; in place maintain a bed of mulch.
Use:	Although the subject has been renowned as an interior tub specimen since the 1880's in Florida, it is often planted in patios and open atriums as a background specimen. The Kentia should never be partly screened, but instead be featured.
Other:	Howea Belmoreana; Similar but more graceful.

Livistonia chinensis

CHINESE FAN PALM

Livistonia: For Livistone, Scotland

chinensis: Chinese

Size:	50 feet, but in Florida usually to 25 feet.
Form:	The 18 inch diameter trunk supports a full round head of fan-shaped leaves 16 feet in diameter.
Texture:	Medium
Leaf:	5 feet wide, palmate flat, the leaf segments split at the middle. Many long tapering ribbons hang vertically below the leaf.
Flower:	Are borne in large numbers in long narrow clusters within the secrecy of the leaves.
Fruit:	Olive-shaped ½ inch long, dull blue.
Geographic Location:	Native to China, Central and South Florida. Hardy to about 22°F.
Dormant:	Evergreen
Culture:	Soil: moderate fertility. Partial shade till maturity fairly slow growth. Positive drainage. Water during dry spells. Spring and summer fertilize. Pests: Scales.
Use:	An old time species. Use as a tubbed subject. In South Florida its use is more often as a landscape subject planted in staggered groups in a triangle similating nature.
Other:	L. chin. subglobosa — a miniature.

Pandanus utilis

SCREW PINE

Pandanus: Latin, Malayan
    name

utilis: Useful

NOTE: SUBJECT IS NOT A
PALM, BUT FITS THE
CATEGORY RATHER
WELL.

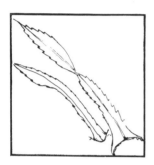

Size:	To 60 feet, but in Florida not over 25 feet.
Form:	Sometimes an erect pyramidal shape, but also an irregular branched open tree with straplike leaf clusters with sharp serrate margins and prop-like roots.
Texture:	Coarse
Leaf:	3 feet long by 3 inches wide; see also above.
Flower:	Inconspicuous in the female. White beard-like bracts fragrant and conspicuous flowing vertically from the flower stalk.
Fruit:	6-8 inch diameter aggregate of tightly pressed fruits of drupes suggesting a pineapple. They turn from a green to yellow as they ripen.
Geographic Location:	Madagascar — South Florida, Tropical, Islands of the Micronesian chain.
Dormant:	Evergreen
Cultures:	Soils can be variable, good drainage and moisture. Sun or dappled shade. Salt tolerance is high. Moderate to fast growth dependant on fertilizer and moisture. Pests: Scale.
Use:	Because of pendant fruits the female is preferred for landscape use. The unusual tree structure makes this tree a favorite for the sculpture effect of near-the-beach landscapes. Young trees 6 feet x 6 feet to 8 feet x 8 feet are favorites for tubs and planter boxes. Somewhat messy because of constant leaf-drop.

12

Phoenix canariensis

CANARY ISLAND DATE PALM

Phoenix: Greek name for Date Palm

canariensis: For Canary Islands

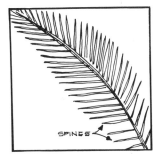

Size:	60 feet, usually up to 30 feet in Florida.
Form:	A hundred pinnate fronds within a crown of 30 feet diameter forms a symmetrical crown growing on top of a massive trunk.
Texture:	Medium to coarse.
Leaf:	Pinnate leaf to 15 feet long, greenish-yellow stems producing lethal lateral spines near base.
Flower:	Unisexual inconspicuous
Fruit:	Ovoid 1 inch in length, Date-like, orange yellow color.
Geographic Location:	Canary Islands, general distribution in Florida. Extra cold hardiness.
Dormant:	Evergreen
Culture:	No apparent special soil and moisture needs. Fertilize twice a year. Fast growing after a slow start prior to forming a trunk. Regular maintenance for trimming fronds. Subject to Date Palm borer, a serious pest.
Use:	Traditional formal street tree of the French Riviera, it has been used as a staggered group of three in the informal manner quite successfully, especially with night lighting. This is a palm for big scale use only.

13

Phoenix reclinata

RECLINATA DATE PALM

Phoenix: Greek name for Date
Palm

reclinata: Latin for reclining

Size:	35 feet
Form:	Graceful curving trunks topped with 12 foot crowns of slender pinnate fronds recurving and dipping to the ground.
Texture:	Medium
Leaf:	Feather-like fronds 7-8 feet long x 3 feet wide. Spines at base of leaf.
Flower:	Insignificant
Fruit:	Oval dates of orange-brown color ¾ inch long, barely edible, viable.
Geographic Location:	Africa. State of Florida to Jacksonville. Below 28°F, damage will occur.
Dormant:	Evergreen
Culture:	Sandy loam, fibrous, good drainage, sunny, fairly wet. Fertilize early Spring and Summer. Rather fast growth. Of easy maintenance. Pests: Scales and leaf spot fungi.
Use:	A big scaled specimen for shade in large open fields. Fairly salt tolerant and wind resistant. An informal subject for use in natural streetscapes. An excellent accent tree for entrances and along patio walks at hotels.

Phoenix roebelenii

PIGMY DATE PALM

Phoenix: Greek for Date Palm

roebelenii: For Mr. Robelen

Size:	9 feet x 6 feet spread of crown.
Form:	Single trunk with plus or minus 70 pinnate leaves gracefully forming a 6 foot diameter crown. Clear trunk to 6 feet.
Texture:	Fine
Leaf:	Pinnate almost lacy to 4 feet long recurving and drooping almost to the soil line. Leaflets slim dark green 3 inch thorns at leaf base.
Flower:	Inconspicuous, 12 inch white sprays in foliage.
Fruit:	Deep red date ½ inch in diameter.
Geographic Location:	Burma, Central and South Florida. Should stand cold damage at 28°F.
Dormant:	Evergreen
Culture:	Any of several fertile soils, good to moderate drainage. Requires moisture. Prefers shade or some sun. Is slow in growth. Added beauty in double and triple trunks. Pests: Florida red scale, brown spot, bud rot.
Use:	Often employed as a terminal accent point in low shrub groups or massing. More formally the Pigmy Date is used as a counterpoint in rythmic linear design with larger material. Is always a favorite in pots, tubs and planter boxes.

15

Ptychosperma elegans

ALEXANDER PALM

Ptychosperma: Greek for grooved seed

elegans: Elegant

Size:	80 feet
Form:	A single ringed trunk, slender, topped by 12 to 15 pendant feathered fronds in a cluster with a spread of crown to 12 feet.
Texture:	Medium
Leaf:	Pinnate fronds 6-8 feet long do not droop below the horizon, leaves are flat and arching, whitish below.
Flower:	Inconspicuous, in much branched spadices below crown 12 inches long.
Fruit:	Drupe-like, bright red, showy round to oval, ½ inch in diameter.
Geographic Location:	Australia. South Florida only, tropical.
Dormant:	Evergreen
Culture:	Full sun to dappled shade, good drainage, tender to any cold weather. Prefers moist conditions for fast growth. Grows best in wet marl. To form a double or triple trunk, two or three seeds are planted in a pot.
Use:	Multiple trunk palms are most popular as patio subjects offering a more natural aspect than the single trunk specimen which is more rigid and formal. Sometimes used on narrow roads and even paths as a streetscape subject planted in line.

Ptychosperma macarthuri

MACARTHUR PALM

Ptychosperma: Greek for
pleated or grooved seed

Size:	30 feet
Form:	Tall thin and many-stemmed, vertical growing cluster to 20 or more feet at which point each stem is crowned by a cluster of feathery flat leaves with each leaflet dark green, broad and with rabbit-nibbled leaf tips.
Texture:	Fine to medium
Leaf:	See under form, size of leaflet 1 foot long x 1-2 inches broad, pinnate.
Flower:	White in bushy clusters, male and female alternating in branched clusters to 15 inches long.
Fruit:	Round, orange-red 3/16 inch in diameter, showy sprays.
Geographic Location:	New Guinea, South Florida only.
Dormant:	Evergreen
Culture:	Subject to damage when exposed to wind. Prefers moist rich soil with two applications of fertilizer. Good drainage. Fairly rapid growing. Sensitive to cold damage.
Use:	Lovely patio subject or as a single accent plant in an atrium. Subject is striking when illuminated. Its strong verticality makes it useful as a specimen at gateways and driveway entrances.

Rhapis excelsa (flabelliformis)

LADY PALM

Rhapis: Greek for needle

excelsa: Tall

Size:	10 feet, often smaller.
Form:	Fan Palms which form bamboo-like clumps with individual canes 1 inch in diameter. Suckering by underground roots.
Texture:	Medium
Leaf:	12 inches wide, fan shaped but deeply divides 5 to 7 segments strap-like 1-2 inches wide.
Flower:	Inconspicuous, in branched spadices hidden by foliage.
Fruit:	Small, 1 to 3 seeded berries.
Geographic Location:	S.E. China, North, South, and Central Florida. Hardy to 20°F.
Dormant:	Evergreen
Culture:	Well drained soil, moist and fertile. Shade to semi-shade recommended. Can resist low levels of light, dust and dryness when used inside. Pests: Scales and caterpillars. Slow growing.
Use:	The shortage of supply has relegated the Lady Palm to a few specialized uses: a) at miniature water falls, b) in oriental settings, c) stylized interior arrangements, and d) potted specimens for Florida rooms and atriums. Its great popularity and demand has made specimens scarce and the price dear.

Roystonia elata

FLORIDA ROYAL PALM

Roystonia: Roy Stone, an American Engineer.

elata: Tall

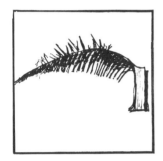

Size:	75 feet
Form:	Columnar feather palm of stately appearance. Trunk solitary; bulged once at the base and once midway up. Topped with a crown of densely grown pinnate leaves 10 feet long.
Texture:	Medium
Leaf:	10 foot long x leaflets growing laterally 20 inches long, many longer.
Flower:	Inconspicuous, in great spadices below the green boot or crown shaft.
Fruit:	Black or bluish drupes about ½ inch long.
Geographic Location:	Native to South Florida.
Dormant:	Evergreen. Withstands frost to 28°F for a limited duration.
Culture:	Rich loam, moderate drainage to swampy. Full sun to broken shade. Native to the S.W. Everglades, propagated by seed. Pests: Scales while young.
Use:	A magnificent boulevard subject, both here as well as in Cuba and Puerto Rico. Today the design preference is in the informal groupings as they are in nature and as seen 25 years ago on the hills outside of Havana.
Other:	Roystonia regina; Cuban Royal Palm, similar.

Sabal palmetto

SABAL PALM

Sabal: Unexplained

palmetto: Little Palm

Size:	80 feet
Form:	Distinctive semi-fan leaved crown atop a slender straight trunk, 12 foot diameter spread of crown.
Texture:	Coarse
Leaf:	Leaves are costate halfway between pinnate and palmate stems are unarmed. Color is gray-green.
Flower:	Inconspicuous, but fragrant and an attractor of bees.
Fruit:	Dull black ⅓ inch in diameter berries. Staple diet for raccoon and gray squirrels in season.
Geographic Location:	Florida to Coastal North Carolina. Cold tolerant.
Dormant:	Evergreen. Native throughout Florida.
Culture:	Amazing adaptability to soils and moisture supply. Sun or shade. Easily transplanted. Fill piled around trunk two feet or ten feet does not seem to effect natural growth. Host to Ficus, Clusia & Brassia plus many orchids, Tillandsias and some ferns.
Use:	If counted by quantities, the Sabal is Florida's most commonly used landscape material. Groups of Sabals are often planted in groups of 20 to 30 varying in height and spacing as well as direction of trees. Seedlings without clear trunks are difficult to transplant. Sabals are always specified by height in clear trunk. Thus, 3 trees in a group might be specified at 1 @ 12 foot C.T., 1 @ 16 foot C.T., and 1 @ 22 foot C.T. (C.T. Clear Trunk)

Serenoa repens

SAW PALMETTO

Serenoa: For Sereno Watson,
Early American Botanist

repens: Latin for creeping,
crawling.

Size:	6 feet x 10 feet
Form:	Sprawling, branching shrub, usually issuing from a subterranean stem. Occasionally with exposed trunk.
Texture:	Medium to fine
Leaf:	Fan-shaped; 4 feet in diameter. It is deeply divided into many radial segments. Petiole armed or not armed.
Flower:	Small, white, massed in elongated branched plume-like clusters, fragrant. Commercial honey is extracted.
Fruit:	¾ inch diameter black berries not palatable.
Geographic Location:	Florida to South Carolina. A native in pine and palmetto woods as well as coastal dunes.
Dormant:	Evergreen
Culture:	No special soil requirements. Medium drainage. Survives on rainfall alone. High shade to full sun where foliage color turns yellow-green. Collected plants from the field should be young to be successful. Very slow growing.
Use:	When using collected stock always double the quantity to anticipate a mortality rate. Lift nature's design for planting by using palmetto as a ground cover.
Other:	A silver leaf strain occurs on parts of the East Coast. This color comes from a waxy coating on the leaves.

21

Trachycarpus fortunei

WINDMILL PALM

Trachycarpus: Greek for rough and harsh fruit

fortunei: Named for R. Fortune, Scottish botanist

Size:	20 feet x 10 feet
Form:	Palmate leaves in crown atop dark trunk usually thicker at the top than at the ground line. Covered with dense hairy fiber.
Texture:	Coarse
Leaf:	Fan-shaped 3 feet in diameter on 18 inch toothed petiole, grey-green in color.
Flower:	Spring, small in panicles 2 feet long; fragrant.
Fruit:	⅓ inch in diameter in drooped spadices 2 feet long among the leaves.
Geographic Location:	Eastern China. Predominantly Central and North Florida. Can withstand 10° F. Will not do well in extreme heat.
Dormant:	Evergreen
Culture:	Soil, fertility above average with good drainage. Fertilize Spring and Summer. Partial shade with wind protection. Watering by irrigation is best. Rapid growth. Pests: Scale
Use:	In early years it makes an attractive picture in selected areas. Again, as a potted or tubbed note it has its place, but of all a 10 foot specimen should be used as an exclamation point or accent where it is stunning.

Washingtonia robusta

MEXICAN WASHINGTON PALM

Washingtonia: Named for George Washington

robusta: Refers to height of tree

Size:	70 feet
Form:	Straight sentry-like trunk with compact, fan-leaf crown usually with a "petticoat" of leaf shag below crown.
Texture:	Coarse
Leaf:	Palmate leaf 3 feet in diameter on petiole, barbed, 3 foot long brilliant green foliage.
Flower:	Inconspicuous, white in branched clusters.
Fruit:	Berry, round and black.
Geographic Location:	Mexico (Sonora & Baja), all of Florida. Hardy to 20°F.
Dormant:	Evergreen
Culture:	Tolerates poor soil and drought, but grows faster with good conditions. The subject was imported into Florida in the 1920s. They came in standing ball to ball and upright in boxcars. All subdivisions were planted along streets, squares, parks and playgrounds.
Use:	This decade, some 60 years later, after the 'boom', nurseries once again are supplying the Mexican Washingtonia Palm. No one recently has dared use it again as a street tree. 'Natural' groupings will appear against hi-rise buildings or in large scale open spaces.
Other:	Specie: W. filifera — not well known in Florida. It is not as clean or as graceful a palm as the one discussed, and has a larger trunk diameter at base.

TREES

Of all the landscape elements used on a building site, the presence of existing trees or even one tree can be the most significant landmark. The design of a proposed residence may be influenced by a tree; it may be the key figure in the design of the garden. The farmer's row of shade trees a generation later may make the most important contribution to the site design of a proposed estate.

Design is the arranging of an entrance arch of grand proportions by bringing a pair of large Oaks some twenty or so feet apart by moving one into position near the existing one and developing an entrance drive between them, shaded by the interwoven branches above.

The form of trees offers the designer at least a baker's dozen of silhouettes to work with from the broad spreading canopy to the vertical column; from the lacy effect of the Parkinsonia to the masculine texture of a Banyan. Each can contribute to the bones of a fine landscape composition.

Herein one will encounter the somewhat amazing intermingling of trees from the temperate zone with those of the tropical zone, thus giving South Florida the largest list of trees in the Western Hemisphere in the zone we call 'subtropical'.

When walking in a high hammock one may encounter a Hickory Tree and the Strangler Rubber Tree growing side by side, or the Carolina Laurel Cherry with its Southern cousin, the West Indies Cherry growing close by.

Acacia auriculiformis

EAR LEAF ACACIA

Acacia: Greek for point

auriculiformis: ear form

Size:	To 40 feet.
Form:	Semi-erect to adulthood when it becomes loose, open and spreading. Brittle wood.
Texture:	Medium
Leaf:	Somewhat sickle-shaped, light bright green, 4 to 6 inches long, parallel veination.
Flower:	Bright yellow
Fruit:	Pod, curling brown thus ear-leafed in form.
Geographic Location:	Australia, Southern Florida
Dormant:	Tolerant 25 to 30°F. Evergreen.
Culture:	Adaptable to soil conditions. Best in full sun. Fast-growing; from 6 to 8 feet per growing season. Stands adverse weather conditions. No pests. Subject to damage in sudden gusts of wind.
Use:	The favorite tree of many land developers for its fantastic growth and rock bottom purchase price. Tolerance to salt conditions can be built into the acacia by early Spring planting (not late Summer or Fall). Almost instant shade tree.

Acer rubrum

RED MAPLE

Acer: Latin for bitter

rubrum: red

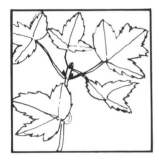

Size:	70 feet x 35 feet spread
Form:	Cylindrical, tall, well-developed trunk and branching habit.
Texture:	Medium
Leaf:	Simple 2½ to 5 inches long, opposite. Deep green above and paler beneath. 3 to 5 lobes.
Flower:	Clusters of dense red at branch tips. Major color note in late Winter.
Fruit:	Pairs of wings, usually red, 1 inch or less long.
Geographic Location:	Native to Florida.
Dormant:	Yes, but late Winter bud-flowers and leafing. Distributed state-wide, very cold-tolerant.
Culture:	Normally wetlands and swampy environmentally, but adaptable to higher drier locations. Dappled shade to sunny exposure. Subject to borers and mites.
Use:	Best known for its color accent when in bloom; attractive red and yellow Fall foliage color. Long-lived. Excellent shade tree. Not brittle.

Amphitecna latifolia

BLACK CALABASH

Amphitecna: obscure

latifolia: broad-leafed

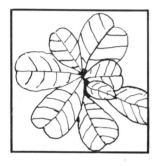

Size:	To 30 feet x 15-20 feet.
Form:	Broad cylinder to rounded crown. Heavily twigged and branched.
Texture:	Medium to coarse.
Leaf:	Alternate, oval to obovate abruptly acuminate 6½ to 10 inches, dark shiny green.
Flower:	Bell-shaped to 2½ inches long. Lavender to yellowish lobes, toothed. Solitary or clustered.
Fruit:	Subglobose 2½ to 3¼ inches long. Thin hard shell, gourd-like.
Geographic Location:	Native: hammocks, Florida peninsula south, Florida Keys, West Indies.
Dormant:	Evergreen, but for a quick drop of old leaves soon replaced with a Spring flush.
Culture:	Native to the limestone substratas and black hammock sands of South Florida. Quite tender to low 30°F readings. Normal moisture requirements. No serious pests.
Use:	Given 8 to 10 years from a sapling, this quick-growing tree develops into a handsome specimen and takes well to shearing and some salt conditions. Its only drawback is its messy leaf-drop before new foliage appears.

Araucaria heterophylla

NORFOLK ISLAND PINE

Araucaria: Arauco, a
  Chilean province

heterophylla: leaves of
  different shapes

Size:	To 100 feet x 18 feet at base, often leaning.
Form:	Vertical, strongly erect pyramid with 5 lateral branches occuring in each tier geometrically.
Texture:	Fine
Leaf:	Mature leaves somewhat triangular and densely overlapping, dark green. Branch interval 1 to 2 feet.
Flower:	Insignificant catkin.
Fruit:	Cones 3 to 6 inches in diameter, almost globular.
Geographic Location:	South Pacific on Norfolk Island. South Florida to Cape Canaveral on coast.
Dormant:	Evergreen
Culture:	Tolerant to a wide range of soils, best with extra moisture and good drainage in full sun. No pests. Absolutely salt-resistant, even in frontal dune areas.
Use:	Wide open spaces such as parks, golf courses in roughs between fairways. Is most effective in mountainous background regions. Do not use in one-story residential areas. Does get wind-damaged in hurricanes.
Other:	A. columnaris (cooki), tall almost fastigiate pyramid.

Bauhinia blakeana

HONG KONG ORCHID TREE

Bauhinia: for the 16th century
herbalists, the
brothers Bauhin

blackeana: for Mr. Blake

Size:	35-40 feet x 30 feet spread.
Form:	Open, spreading. Canopy shade tree.
Texture:	Medium to coarse.
Leaf:	Bipartite foot-print to 8 inches wide. Light green.
Flower:	Orchid-like to 6 inches wide, rose purple in sprays on outer branches from Fall to Spring.
Fruit:	Non-producing.
Geographic Location:	Southern China
Dormant:	Evergreen, but semi-deciduous when in bloom.
Culture:	Tolerant as to soils, 80% sunny or more. Normal moist to semi-moist conditions but should be well-drained. Damage to 28°F.
Use:	Where Winter-season color is required Blakeana fills a sparse bill. Has been used in streetscapes with some success. Otherwise a shade tree of some note in natural settings. A rapid-grower.
Other:	B. purpurea: to 40 feet, deciduous, orchid flowers to 5 inch diameter. B. purpurea candida: a pure white-flowered variety.

Bischofia javanica

JAVA BISHOPWOOD

Bischofia: taken from G.
     Bishoff, German botanist

javanica: Javanese

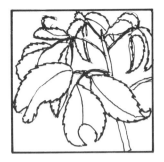

Size:	60 feet x 35 to 45 feet.
Form:	Dense, rounded, haystack head from a huge trunk.
Texture:	Medium to coarse.
Leaf:	Alternate, compound, 3 oval leaflets to 8 inches pointed and finely serrated.
Flower:	Dioecious, small greenish-yellow, inconspicuous.
Fruit:	Round, ⅓ inch in diameter, fleshy with 3 to 6 seeds.
Geographic Location:	Tropical Asia and Pacific Islands. South Florida and coastal to Sarasota and Canaveral.
Dormant:	Evergreen. Frost-sensitive at 28°F.
Culture:	Fast-grower in full sun on sandy soils, but often grown on marshland and muck. Foliage subject to sooty mold.
Use:	Probably overused on small residences as it soon outgrows in scale. Its dense crown inhibits the growth of grass. Better used on large sites as on parklands and institutional sites. Somewhat subject to wind damage. A dense shade tree.

Bucida buceras

BLACK OLIVE

Bucida: Latin for crooked
    horn, refers to
    seed capsule

buceras: ox-horned

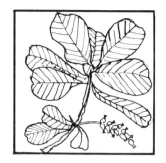

Size:	60-80 feet x 25 feet.
Form:	Semi-pyramidal in juvenile stage to spreading and rounded as an adult.
Texture:	Medium
Leaf:	Size greatly variable to 3½ inches long; clustered at twig ends. Alternate, shiny.
Flower:	Inconspicuous
Fruit:	Small bottle-shaped capsule, occasionally viable seed.
Geographic Location:	Greater Antilles and Leeward Islands. South Florida only.
Dormant:	Evergreen; tender at 32°F. Can be killed at 25°F.
Culture:	Natural environment; brackish wetlands with buttonwoods and white mangroves but adaptable to Florida soils; prefers sun with moisture. Pests: sooty mold and bark borer.
Use:	Probably our most popular tree for street plantings as well as for home use. Caution is advised in use near automobiles, concrete decks and sidewalks as well as white tile roofs; the seed capsules exude a tannic acid stain which is long-lasting. Excells in streetscapes and median strips.

Bucida 'Shady Lady'

Bucida: Crooked horn, refers
to seed capsule

Var: Shady Lady

Size:	25 feet x 16 feet.
Form:	Compact, semi-cylindrical. Remarkably symmetrical and identical. Rather formal habit of growth.
Texture:	Fine
Leaf:	1½ x ½ inch,obovate at the tips, tapered to base. Needle - like spines ½ to ¾ inch long. Most joints branch in pairs. 3 to 5 leaves at joints.
Flower:	Not seen.
Fruit:	Not seen.
Geographic Location:	Developed in Southern Florida. Leaf damage at 28°F.
Dormant:	Evergreen.
Culture:	Somewhat more resistant to salt exposure than B. buceras and similar to B. spinosa. Tolerant as to wet and dry conditions as well as various soils. Thrives on fertilizer, sunny locations and good drainage.
Use:	Excellent as a vertical note in the natural landscape as well as a singularly forceful effect as reiterative element in linear design. Proving a happy choice in streetscapes.
Other:	B. spinosa, native to Bahamas, spurs developed as thorns, often with picturesque irregular form. 25 feet x 16 feet.

Bursera simaruba

GUMBO LIMBO

Bursera: for Joachim Burser, a
European botanist

simaruba: resembling the
Paradise tree, Simara

Size:	75 feet x 35 feet
Form:	Juvenile semi-upright; broad-spreading adult.
Texture:	Medium
Leaf:	Compound, alternate pinnate 6-8 inches long, one terminal.
Flower:	Inconspicuous
Fruit:	Clusters, dark red. Elliptic ½ inch in diameter.
Geographic Location:	Bahamas, Florida, Central America, Caribbean.
Dormant:	Only in cold years. Otherwise partial defoliation.
Culture:	Thrives with neglect. Fast-growing. Brittle branchwork. Soil-tolerant. Full sun or shade. No pests.
Use:	Known for its summertime shade. Exfoliating red bark on trunk and branches easily identify it. Native in hammocks. Haitian voodoo drums called tambours are carved from trunks. Often of giant diameters. Strongly dominant in most hammocks.

Calophyllum antillanum

CALABA, BEAUTY LEAF,
MARIA IN CUBA

Calophyllum: Greek for
beautiful leaf

antillanum: from the Upper
West Indies, the Antilles.

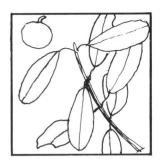

Size:	80 feet x 15-20 feet.
Form:	Semi-pyramidal to wide columnar.
Texture:	Medium to coarse.
Leaf:	Ovate to 6 inches long. Pinkish tips on new growth, opposite, leathery.
Flower:	White, fragrant, in axillary few-flowered racemes.
Fruit:	Brown, globose 1 inch in diameter.
Geographic Location:	West Indies.
Dormant:	Evergreen
Culture:	Found in dense woodlands, in hills and wetlands. Can withstand innundation and is resistant to brackish conditions. Root-pruned trees to 40 feet transplant easily. No pests.
Use:	One of the handsome trees of the zone II coastal area. It combines well in a mixed planting of seagrape, mimusops and calophyllum.

Callophyllum inophyllum

ALEXANDER LAUREL
OR KAMANI

Callophyllum: Latin referring
to its beautiful leaves

inophyllum: translated in
the leaf

Size:	60 feet
Form:	Ascending branches, rather norrow, ovate-shaped tree.
Texture:	Medium to coarse
Leaf:	Shiny, leathery, nearly oval, 3 to 8 inches with many fine, parallel, diagonal side veins, blunt tips.
Flower:	White, 1 inch in diameter, fragrant in clusters of four to fifteen.
Fruit:	Globose, green, 1 inch wide, clustered thin skin covering a bony shell which holds the seed.
Geographic Location:	Coastal India and Southwest Pacific, South Florida, coastal to Palm Beach.
Dormant:	Evergreen
Culture:	Slow-growing, resistant to wind and salt, full sun, tendency to lean. This is a subtropical and is not cold-hardy. Pests: None.
Use:	Excellent high screening as a tree especially in the salt zone. Individual specimens are valued for their handsome shiny green foliage and make good shade trees. Counted as one of 6 or 8 most salt-tolerant trees.

Callistemon lanceolata

LEMON BOTTLEBRUSH

Callistemon: translates to beautiful stamens in the flower.

lanceolata: lance-shaped leaves

Size:	12-20 feet x 8-10 feet
Form:	Wide cylinder.
Texture:	Fine to medium.
Leaf:	Stiff, narrow leaves 1 to 3 inches by ¼ inch vein arrangements, pinnate.
Flower:	Showy red flower spikes 2 to 4 inches long by 2½ inches wide, born upright near branch ends.
Fruit:	Capsule globose, contracted at tip.
Geographic Location:	Australia
Dormant:	Evergreen, withstands light frost in central Florida better than Callistemon viminallis.
Culture:	Tolerant to wetland conditions as well as dry, sandy soils. Full sun, fast-growing. Fertilize 2 to 3 times per year. Pests: reasonably immune.
Use:	A line of these on the west side cast stunning shadows in a repetitive pattern. As a screen or buffer, excellent as it suckers at the base. Is best used as an occasional color note in a mixed planting. Somewhat difficult to transplant, B and B.

Callistemon viminallis

WEEPING BOTTLEBRUSH

Callistemon: translates to
beautiful stamens, refers
to flower.

*viminallis:*leaves for use in
wickerwork, willowy

Size:	20 feet x 20 feet.
Form:	Rounded head with strong weeping habit.
Texture:	Fine
Leaf:	Alternate, slender to 4 inches long, tomentose when young.
Flower:	Numerous deep red stamens in long pendant cylindrical spikes 1½ inches wide.
Fruit:	Capsules containing tiny papery seeds
Geographic Location:	Australia
Dormant:	Evergreen, tender to cold conditions.
Culture:	Can thrive in wet-bottom land situations. Grows well in dry sandy soils.
Use:	Two major attributes: famous for its twice-a-year blooming period, and is ideal for small home plantings because of its limited height and spread. Character and form can be developed by judicious pruning to develop an oriental form. Useful on pond margins and lake shores for its willow-like habit of growth.

Cassia surattensis

GLAUCOUS CASSIA

Cassia: Greek ancient name

surattensis: geographic source
in India

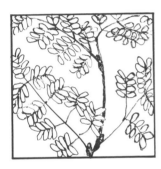

Size:	25 feet x 10 feet wide.
Form:	Upright to slightly spreading in adulthood.
Texture:	Fine
Leaf:	Compound with 6 to 10 blunt oval leaflets to 2 inches long.
Flower:	2½ inches wide in short spikes, golden yellow occuring in Spring and Fall. Blooms twice a year.
Fruit:	Seed pods 4 to 6 inches long x 2½ inches wide exploding when ripe.
Geographic Location:	Southeast Asia, Australia, South Pacific.
Dormant:	None, can withstand 30°F temperature.
Culture:	Soil-tolerant, prefers full sun. Short-lived but fast-growing, average moisture. Reseeds itself with ease. Pests: few.
Use:	Is valued as a dooryard tree of small stature. Its major handicap seems to be a rather weak root structure which does not resist strong winds. In a land of giant and big trees this tree is gaining popularity for its medium size.

Cecropia peltata

TRUMPET TREE,
SNAKEWOOD

Cecropia: refers to wood used
in woodwind instruments

peltata: peltate as in a
stretched skin

Size:	To 50 feet x 15-20 foot spread.
Form:	Umbrella top with open branchwork with a cluster of large leaves at the terminus of the branches.
Texture:	Coarse
Leaf:	12 inches wide, green on top, silvery-white underneath. Divided ⅓ deep to center with 7 to 11 lobes.
Flower:	In slender catkins.
Fruit:	Slender catkins, fleshy with many small seeds, greenish to brown at maturity.
Geographic Location:	Tropical America, Caribbean Islands.
Dormant:	Evergreen. Cannot withstand frost damage without a setback.
Culture:	Fast-growing, short-lived, found in moist soils in ravines protected from harsh winds. Also a rain-forest inhabitant. Starts in part-shade, but soon grows to the top of surrounding trees.
Use:	Summer breezes shift the foliage suddenly from green to silver as the leaves are reversed. Suited for use in the mixed forest and in colonies where shade is useful. Special effects such as ravine plantings and hillside plantings.
Other:	C. palmata. A less impressive tree but more common.

Celtis laevigata

HACKBERRY, SUGARBERRY

Celtis: obscure

laevigata: smooth and
   polished, refers to bark
   in old trees

Size:	60 feet to 80 feet, trunks 2 to 3 feet in diameter.
Form:	Wide-spreading with round top. Smooth trunk peppered with warts. Branches open and spreading, tips pendulous.
Texture:	Medium
Leaf:	Simple, alternate, thin 2½ to 5 inches long, narrowly ovate with long sharp tip, unequally rounded or wedge bases.
Flower:	Inconspicuous, late Spring, stalked, greenish at axis of new shoots.
Fruit:	Orange-red to yellow drupe ¼ inch in diameter with a wrinkled bony pit. Distributed by birds.
Geographic Location:	Distributed over most of Florida as far as the Florida Keys, West Indies and Mexico. Hammocks and river bottoms.
Dormant:	Deciduous
Culture:	Moist forest loam, also clay and silt-type soils. Sun or shade. Found in colonies or mixed hardwoods. Growth is rapid. No serious pests.
Use:	Celtis is an attractive shade tree for homegrounds. Its light and airy foliage gives the tree contrast to the dominant dark green foliage seen in Florida. A must for projects involving the native plant palette.

Chrysophyllum oliviforme

SATIN LEAF

Chrysophyllum: Greek for
golden leaf

oliviforme: olive-shaped

Size:	30 feet x 15-20 feet.
Form:	Usually haystack-shaped, occasionally irregular.
Texture:	Medium
Leaf:	Shiny green above, lustrous copper pubescence beneath, 2 to 4 inches long.
Flower:	Clustered
Fruit:	Dark purple olive-shaped, about ¾ inch long. Single seed.
Geographic Location:	Native, South Florida hammocks, West Indies. Found on the coast to Merritt Island.
Dormant:	Evergreen
Culture:	Soils: moist marl to sandy loam. Withstands forest shade until mature tree. Medium to fast growth. Subject to leaf-indentured scale.
Use:	A valuable native. Foliage becomes bicolored when breezes reverse leaves. A symmetrical specimen canopy tree in open lawns with age. Withstands light frost. Framework is upright then spreading. Among the ten most important natives.

Citrus aurantifolia

KEY LIME

Citrus: ancient classical name

aurantifolia: golden leaf

Size:	12-14 feet x 8 feet.
Form:	Upright to round, sometimes multiple-trunked.
Texture:	Medium
Leaf:	Yellow-green to green, to 3 inches long. Pungent spines on branches are plentiful.
Flower:	6 petals, white 3/8 inch diameter at base of leaf.
Fruit:	Round to plump oval, green to yellow when mature, several seasons, numerous seeds. Thin skin.
Geographic Location:	Origin Asia. Now Spain, South America, Mexico, Caribbean, Bahamas.
Dormant:	Evergreen. Tropical only.
Culture:	Usually sandy loam, well-drained, fertilize sparingly, full sun.
Use:	Although not the classic form, its popularity is based on its very well-known fruit used in desserts, cookery and drinks. Trees are usually planted in rear or non-use areas. Short-lived.
Other:	Spineless Key Lime: fruit globose, smaller, yellow. C. auranti, Taitensis, Tahiti Lime: larger fruit, green. C. Meyer: large globose fruit, sweet-sour, semi-bushy to 8 feet.

Citrofortunella mitis

CALOMONDINE

Citrus: ancient name

fortunella: for R. Fortune,
  English traveller

mitis: Latin for mild, mellow

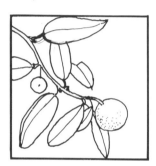

Size:	To 25 feet, usually less.
Form:	Upright, columnar, dense crown.
Texture:	Fine.
Leaf:	1-3 inches long, slightly scalloped, some spines on branches.
Flower:	Small multitudes of white citrus flowers, fragrant.
Fruit:	Flattened, globose to 1 inch diameter. Deep orange colored, sour and few seeds. used as an "ade".
Geographic Location:	Universal in the world's tropics. Origin was Asia, then Asia Minor.
Dormant:	Evergreen. Withstands light frost (28°F).
Cultutre:	In sandy soil add peat, shavings and mulch to 12 inches deep. Well-drained conditions only. Citrus needs moisture. Groves are irrigated. Fertilize in March, June and September. Pests: scale, virus, mealy bugs.
Use:	Primary use as an ornamental; not a fruit to eat fresh. Excellent foliage and fruit color. Compact tree for use as an accent in home design. Also can be planted formally in a linear design spaced on centers of 10 to 18 feet.

Citrus paradisi

GRAPEFRUIT

Citrus: ancient classical name

paradisi: like paradise

Size:	To 30 feet x 25 feet spread.
Form:	Dense, rounded crown. Trunk usually less than 4 feet clear of grade.
Texture:	Medium to coarse.
Leaf:	To 6 inches long, dark green, leathery and shiny.
Flower:	White to 1½ inches wide in clusters, fragrant.
Fruit:	Globose, flattened globose or pear-shaped to 5 inches wide, yellow skin. Usually seedless, but use Marsh Seedless to be sure.
Geographic Location:	Originated in the West Indies as a sport. Introduced to Florida about 1809. Parent probably was C. pummelo.
Dormant:	Evergreen, but very cold- or freeze-resistant, best grown from Vero Beach South.
Culture:	Ideal in rich sandy loam with plenty of moisture, fertilize 3 times per year. As a grove tree, use irrigation.
Use:	Not an ornamental, but handsome as a fruit tree and can form a background screen blocking unwanted views well.
Other:	Marsh Seedless: large, late, white meat. Ruby red: very popular in Rio Grande, Texas and well adapted to South Florida.

Citrus sinensis

SWEET ORANGE

Citrus: ancient classical name

sinensis: Latin for
   Chinese origin

Size:	To 40 feet x 25 feet spread.
Form:	Densely branched, conical or rounded head. Sometimes a few thorns.
Texture:	Medium
Leaf:	Dark, shiny green, oval to 4 inches long.
Flower:	White, 1 inch or more in width, very fragrant.
Fruit:	Round up to 4 inches in diameter, yellow to golden orange, singly or clustered.
Geographic Location:	China
Dormant:	Evergreen, tolerant to 26°F for a few hours.
Culture:	Rich, sandy loam, well-drained but moist. Full sun. Fertilize in March, June and September. Must be budded or grafted. Pests: leaf canker, scale, white fly.
Use:	This is the grove-type or commercial orange. One or two would be ample for home owners' use. Blends well with other tropical trees or is a good dooryard specimen for flanking views.
Other:	Valencia, Temple, Pineapple, Parson Brown and Navel.

Clusia guttifera

SMALL-LEAFED CLUSIA

Clusia: an association of
   wetland places

guttifera: sticky, referring
   to sap.

Size:	18-25 feet x equal spread.
Form:	Early form is pyramidal. Adult round and spreading, compact rather than open as in C. rosea.
Texture:	Coarse to medium.
Leaf:	Leathery, pear-shaped, tapering at base, 5 to 6 inches long.
Flower:	Pink, waxy, camelia-like, prominent button. Center is sticky.
Fruit:	In Spring or early Summer, globular, 8 parted breaking from top, curving back to expose sticky seeds to birds.
Geographic Location:	Caribbean
Dormant:	Evergreen. Severe damage below freezing.
Culture:	Rich moist loam, but well-drained. Tolerant to some shade or full sun. No pests.
Use:	Has surpassed C. rosea in usage because of compactness and less coarse in texture. Use as a specimen in open areas or as a soft buffer planting or background. Very adaptable for espalier effects but not on hot exposed walls.

Clusia Rosea

PITCH APPLE,
  OR CARIBBEAN
  MONKEY APPLE

Clusia: from an association
  from wet or
  flooded places

rosea: derived from pink-
  colored flower petals

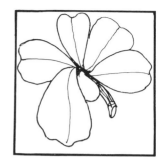

Size:	20 to 50 feet x 40 feet.
Form:	Spreading round head.
Texture:	Coarse
Leaf:	Opposite, short-petioled, rounded obovate and narrowed at base, 8 inches x 4½ inches wide, thick, leathery.
Flower:	Showy, 3 to 4 inches wide, face of petals pink or white, 6 petals camelia-like.
Fruit:	Capsule 3 inches wide opening to present some red fleshed seeds. Loved as bird food.
Geographic Location:	West Indies, Bahamas, hammocks and moist ravines.
Dormant:	Evergreen. Severely burned at freezing temperatures.
Culture:	Sandy loam, sunny to dappled shade, moist soil condition. Fertilizer twice per Summer. Rapid growth.
Use:	Superb espalier subject on cool walls. As a small to medium tree acts as an excellent screen or buffer against commercial areas because of its low-spreading habit. Also valued for its handsome dark green foliage.
Other:	C. rosea variegata: its yellow and green foliage produces a showy tree.

48

Coccoloba diversifolia

Pidgeon Plum

Coccoloba: Greek for
lobed berry

diversifolia: varied leaf form

Size:	20 to 50 feet x 35 feet.
Form:	Spreading round head. Compact with a density of foliage.
Texture:	Medium to coarse.
Leaf:	Alternate, 3-4 inches long, ovate-oblong tips, round or pointed. Leathery, shiny surface.
Flower:	Spikes terminal or axillary. Small, numerous.
Fruit:	Berry-like ⅓ inch diameter. Black in 3 inch long racemes.
Geographic Location:	South Florida, Bahamas and West Indies. Hardwood hammocks.
Dormant:	Evergreen except for quick change in March.
Culture:	Semi-shade to sunny. Sandy loam, moist condition. Fast-growing, good drainage. No serious pests.
Use:	Streetscapes and median strips. Popular as a residential subject for uniform form, texture and leaf drop. For uniformity use as a single clone for vegetative reproduction in place of other methods. Berries attractive to birds.

Coccoloba uvifera

SEAGRAPE

Coccoloba: Greek for
   lobed berry

uvifera: Greek for
   grape-bearing

Size:	30 feet x 25 feet.
Form:	Spreading shrub on sand dunes to irregular, spreading tree away from dune line. Can be picturesque.
Texture:	Very coarse.
Leaf:	Rounded, notched at base, radially veined 5 to 8 inch diameter, dark green, leathery.
Flower:	Small flowers in catkin-like spikes.
Fruit:	¾ inch diameter, pear-shaped, rosy in "grape" clusters in late Summer. Famous as seagrape jelly.
Geographic Location:	Caribbean, Southern Florida and Bahamas, back of tidal level dunes and hammocks.
Dormant:	Evergreen. Short period of leaf change in late March.
Culture:	Sunny, moist, sandy and well-drained. Rank growing. Adapts well to pruning. Salt-resisting. Pests: twigs subject to seagrape borer.
Use:	Has succeeded as a ground cover and erosion control on 3 to 1 hillsides. As an accent tree with frame residences. Forms contorted, stratified forms for canopy trees with great character when shaped. Good multiple-trunked. Fruit most attractive to birds and squirrels.

Conocarpus erectus

BUTTONWOOD

Conocarpus: Latin for
cone-shaped fruit

erectus: Latin for erect,
referring to "cone"

Size:	60 feet x 40 feet.
Form:	Low-branching, shrubby, rounded crown, spreading.
Texture:	Medium to fine.
Leaf:	1 to 4 inches long, alternate, elliptic with pointed end. At base of blade, above the petiole, see 2 small glands.
Flower:	Minute in cone-like clusters.
Fruit:	Fruiting heads mature into cone-like structures.
Geographic Location:	South Florida at rear (landward) of mangroves and on fringes of low hammocks. Caribbean, Bahamas.
Dormant:	Evergreen
Culture:	Highly salt-tolerant. Sunny locations. Sandy or marl soils, good moisture, but not inundation. Responds to fertilizers. Pests: scale, sooty mould, mites.
Use:	Adapts well to hedge use and shearing in lieu of Autralian pine. Especially useful near coastal beaches; zones I and II. Shrubby character makes it a useful background and screen plant. Prime salt-tolerance.
Other:	C. erectus var. sericeus. Silver-leaf buttonwood is more popular than its green parent.

Cordia boissieri
ANACAHUITA,
   WHITE CORDIA
Cordia: refers to
   heart-shaped leaves
boissieri: probably for Pierre
   Boissier, Frenchman
   1810-1885
Note: this tree has been
reported as close to extinction.

Size:	To 20 feet tall and 6 to 8 inch caliper
Form:	Somewhat broad-pyramidal
Texture:	Medium
Leaf:	To 5 inches long, velvety tomentose by 4 inches wide, margin crenulate to entire. Simple, dark green.
Flower:	In terminal clusters, 1½ inch long, white with yellow centers, striking.
Fruit:	One-seeded, yellow drupe, ½ inch long, bright red-brown seed. Fleshy white exterior
Geographic Location:	Lower Rio Grande Valley, Texas, South New Mexico.
Dormant:	Evergreen
Culture:	Sunny exposure. Reacts pleasantly to sandy loam, plenty of moisture and one or two fertilizings per growing season resulting in profuse flowering and seed production. Pests: unknown.
Use:	This small tree has withstood 25°F. As a single specimen in an open lawn it deserves much more public acclaim. Presently not well-known.

Cordia sebestena

GEIGER TREE

Cordia: for V. Cordus, 16th
century German botanist

sebestena: an Arabic name for
Geiger tree, named after
John Geiger, Key West
pilot and wrecker

Size:	30 feet x 15 feet.
Form:	Dense, rounded crown when mature.
Texture:	Medium to coarse.
Leaf:	Rough, dark green, stiff, oval 3 to 5 inches long, sand-papery surface.
Flower:	Orange to scarlet flowers in cymes, 1-2 inches long, geranium-like.
Fruit:	Drupe, white ¾ inch long "snowberry".
Geographic Location:	Native, South Florida, Bahamas, Caribbean, South America.
Dormant:	None. Not cold-hardy.
Culture:	Sandy soils under dry conditions, full sun, close to zone I, salt-resistant
Use:	One of best small broad-leafed trees for salt-resistance. Valued for its smashing Spring and Summer blooms. Well adapted for natural plantings in parks and small homes near the ocean or Gulf of Mexico in tropic Florida.
Other:	C. boissiere: a smallish Geiger from the Rio Grande Valley with stunning white flowers in clusters. Valued for its frost-resistance.

Cupianopsis anacardioides

CARROTWOOD

Cupianopsis: cup-shaped or resembling cup

anacardioides: Greek for without heart

Size:	30 x 30 feet
Form:	Umbrella canopy, single trunk, compact.
Texture:	Medium
Leaf:	Compound, 6-10 leaflets, leathery. Each 4 inches long, handsomely arranged.
Flower:	Axillary raceme 10 inches long. Many small lime-green flowers each 3/8 inch in diameter.
Fruit:	Branched raceme carrying many globose, ridged fruits. Seeds within, one in 3 cells. Fruit ½ inch in diameter.
Geographic Location:	Australia
Dormant:	Evergreen
Culture:	Tolerates poor soil. Prune errant branching to clean up form. Average watering. Poor drainage tolerance is good. Slow to moderate growth. Old trees tolerate 22°F with little damage. Sunny.
Use:	Clean tree for poolside use. Gives heavy, dense shade. Handsome patio, lawn or street tree. Deep-rooting. Fairly salt-resistant. Adapts to dry areas. Rather recent introduction into Florida, but well known in California.

54

Dalbergia sissoo

INDIAN ROSEWOOD, SHISHUM

Dalbergia: for the physician Nils
Dalberg of Sweden

Sissoo: an old Indian name

Size:	To 45 feet x 25 feet.
Form:	Upright to rounded, compact, symmetrical crown.
Texture:	Medium
Leaf:	Compound pinnate of 3 to 5 oval pointed leaflets, alternate, each 1 to 3 inches long.
Flower:	Pinnacles of small yellowish-white, short-stemmed flowers occuring in leaf axils and branch tips.
Fruit:	Smooth, thin, light brown pods to 4 inches x ½ inch wide. 1 to 4 flat seeds.
Geographic Location:	India, South Florida to Cape Canaveral. On Gulf coast to Sarasota.
Dormant:	Semi-deciduous. New foliage by March 15th.
Culture:	Estimated cold-tolerance is 25°F for a limited time, about 4 hours. Best in sun. Can grow equally well in dry conditions as well as moist. Fertilize twice per growing season. Pests: none serious.
Use:	When limbed up to 6 feet the rosewood can be used for easy sight lines under canopies when grouped. Good shade tree during warm season. Fast growth.

Delonix regia

ROYAL POINCIANA,
FLAMBOYANT

Delonix: Greek, refers to
long-clawed petals

regia: Latin for royal

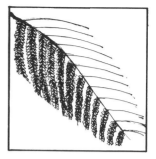

Size:	30 feet x 50 feet
Form:	Broad haystack crown to flat-topped. Rather stratified branchwork.
Texture:	Fine
Leaf:	Compound leaf to 2 feet long, pinnate with tiny leaflets.
Flower:	Showy clusters, 4 inches wide forming a blanket of bright orange color over top and sides of crown. June and July.
Fruit:	Strap-like pods, brown, 2 feet long, 2 inches wide, many seeds elongated.
Geographic Location:	Madagascar, Africa.
Dormant:	Deciduous, November and May.
Culture:	Very rapid grower, almost any soil, with minimum water. Sunny, well-drained condition. Sensitive to freeze damage. Pests: none serious.
Use:	As a specimen shade tree for open lawns. As a street tree (even spacing) on wide boulevards. Can be intermixed to absorb the long dormant period. Surely our most spectacular flowering tree. A non-pareil for large open spaces.
Other:	A yellow flowered (golden) variety from Tobago and also Jamaica has been introduced.

Diospyros ebenaster

BLACK SAPOTE

Diospyros: twisted in the air

*ebenaster:*black and star-like

Size:	To 45 feet
Form:	Beautifully rounded crown of dense foliage.
Texture:	Medium to coarse
Leaf:	Smooth blunt leaves 4 to 8 inches long by 2½ to 3 inches broad.
Flower:	Inconspicuous, yellow-green, alternate, simple.
Fruit:	Green to black, globose, about 4 inches in diameter, tomatoe-shaped, insipid.
Geographic Location:	Mexico and West Indies. Tropic areas in South Florida.
Dormant:	Evergreen. Can tolerate only 2 or 3°F of frost.
Culture:	Fast-growing, semi-shade to full sun. Average moisture, good drainage. Prefers rich organic soil. Resistant to most pests.
Use:	Considered a valued tree for its handsome thick dark green canopy and shade characteristics. Its only drawback is the windfall of the sticky fruit.

Dodonea viscosa

VARNISH LEAF

Dodonea: for the Dutch botanist
Rembert Dodoens

viscosa: from Latin for the sticky
resinous surface of the leaf

Size:	10 to 15 feet
Form:	Loosely rounded and rather open.
Texture:	Medium
Leaf:	Narrowly oblanceolate 2¼ to 3¼ inches long x ½-¼ inch wide, with lacquered upper surface.
Flower:	Yellow-green to reddish, appears in clustered capsules. Terminal, small, inconspicuous.
Fruit:	Three-celled seed capsules with 3 or 4 rounded wings with a red or purple infusion. Capsules light green.
Geographic Location:	India, South Florida in hammocks. Keys and West Indies. Native.
Dormant:	Evergreen
Culture:	D. viscosa is both wind- and salt-resistant, tolerant of drought conditions, sun or partial shade. Young plants are easily moved. No serious pests. Often found thick in disturbed areas.
Use:	Little experimenting has been done by designers in Florida. However, Richard Workman in *"Growing Native"* points to the West Indies where Dodonea is planted in hedges and grown in fence rows.
Other:	D. v. purpurea: a color break in foliage from California.

Eriobotrya japonica

LOQUAT

Eriobotrya: Greek, "erio" for
 wooly, "botrya" for cluster.

japonica: Latin for Japan

Size:	10 to 20 feet x 16-20 feet or more in sun. Slender in shade.
Form:	When limbed up this is a handsome tree usually classed as a small to medium. Rounded crown.
Texture:	Coarse.
Leaf:	Long, oval, shiny above and rough pubescens below. Handsome and about 10 inches long, toothed edges.
Flower:	Fragrant, white, small and formed on wooly bracked spikes in Spring.
Fruit:	In clusters 1-2 inches long, oval, with slight neck, apricot yellow, furry. Delicious flesh, subacid, brown.
Geographic Location:	Japan. Cold-hardy to North Carolina. China.
Dormant:	Evergreen
Culture:	Sunny, well-drained. Average soil conditions. When established can stand drought conditions. Selected varieties, when grafted, are best for improved fruit. Pests: mites, scales, Med fly and caterpillars.
Use:	Select shapes for espalier work on cool walls. Common as a dooryard fruit and ornamental. Fruit attractive to birds. Ideal for residential scale use.
Other:	E. j. 'Deflexa': leaves a bright coppery red.

Eucalyptus torelliana

TORELL'S EUCALYPTUS

Eucalyptus: Greek for
well-covered.

torelliana: after Torell

Size:	30 to 35 feet x 15-20 feet.
Form:	Broad, cylinder, loose-branching, not stiff.
Texture:	Medium to coarse
Leaf:	New leaves on terminal shoots are fuzzy rouge red, 4-5 inches x 3 inches, ovate with acute tip. Rather thin but with a rough paper surface on both sides.
Flower:	Small, white, multi-filament, in clusters of 3 each in panicles.
Fruit:	Capsule, brown, ¾ inch long, in clusters of threes on branched panicles, terminal of 12-15 inches long.
Geographic Location:	Australia
Dormant:	Evergreen. Considerable freeze damage at 25°F.
Culture:	Very fast growing. Guard against tree roots that are ring-grown (circles in cans). Sandy loam. Use rotted cow manure for backfilling in pits. Sunny, well-drained. No pests.
Use:	Useful addition to the landscape scene. Subject to winds 20-30 mph. Therefore, plant in triangles 10 feet or more on a leg for wind protection. Makes a beautiful screen for background and buffers. Handsome, smooth, dappled trunk. Some salt tolerance.

Eugenia axillaris

WHITE STOPPER

Eugenia: Latin for Prince Eugene of Savoy, patron of botany

axillaris: Latin for axillary flower arrangement

Size:	25 feet x 9 feet.
Form:	Cylindrical to erect.
Texture:	Medium
Leaf:	Opposite, simple, leathery, oval to elliptic, 3 x 1 inch.
Flower:	Rather inconspicuous, axillary raceme with 5 to 7 pairs of flowers.
Fruit:	Berry ½ inch in diameter, green to red to black in clusters of 3 to 5. Juicy, sweet.
Geographic Location:	Native, North to St. John's River to Key West, in hammocks. West Indies, Central America.
Dormant:	Evergreen
Culture:	Easily grown from seed. Fast grower. In sun it forms thickets. In shade it thrives and grows fast as individual specimen trees. Not hard to transplant. No serious pests.
Use:	Very desirable for tall hedges or informal screens and is an excellent tubbed plant. Fairly salt-tolerant. Exudes a pungent odor. Background planting in mixed natural plantings. An understory tree, a strong link between canopy and low shrubs and ground cover.

61

Eugenia foetida

SPANISH STOPPER OR
BOX-LEAF EUGENIA

Eugenia: for Prince Eugene of
Savoy, patron of botany

foetida: for stinking or
evil-smelling

Size:	20-35 feet, trunks to 8 inch diameter.
Form:	Compact, erect. Old plants may be picturesque and irregular in branchwork.
Texture:	Fine
Leaf:	Leathery, opposite, dark green and glossy above with curled under-margins, 1½-2¼ inches long. Obovate.
Flower:	White, tiny with 5-7 pairs of flowers on short axillary raceme.
Fruit:	Globose, dark red to black, ¼ inch wide with 1 or 2 seeds.
Geographic Location:	A native in hammocks near coast, South Florida, Central America, Bahamas and Caribbean.
Dormant:	Evergreen
Culture:	Happy in rich, sandy loam in dappled sunlight, reasonably moist ground without irrigation. Pests: none serious.
Use:	An understory tree of handsome form which can add an extra dimension between upper canopy and low shrubs. Useful as a tall hedge or as a tubbed specimen. Heavy crops of berries are a source of food for birds.
Other:	Lesser known species are: E. simpsoni — Simpson's stopper. E. confusa — Redberry stopper. E. longipes — Long-stalked stopper. E. rhombea — Red stopper.

Exothea paniculata

INKWOOD, BUTTERBOUGH

Exothea: Latin for outer wall
   of the anther

paniculata: flowers panicled

Size:	35 feet x 20 feet.
Form:	Upright and slender. Starts as a shrub in hammocks.
Texture:	Medium
Leaf:	Alternately compound with 2 to 6 (usually 4) leaflets, dark green, shining, 2 to 5 inches long, elliptic.
Flower:	In Spring male and female on separate trees, small, white flowers clustered at ends of branches.
Fruit:	Subglobose, glabrous at branch ends, about ¼ inch in diameter. Orange turning purple.
Geographic Location:	Florida Keys, northward to Brevard County, along the Central America and West Indies.
Dormant:	Evergreen
Culture:	Tolerates some cold (to 27°F). Hammock conditions for usage. Dappled shade and well-drained soils.
Use:	Best used as tree for mixed hardwoods and hammocks. Excellent shade tree characteristics. As native plants become popular, look for better availability.

Ficus aurea

STRANGLER FIG

Ficus: Latin for fig

aurea: Latin for golden

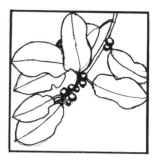

Size:	50 feet x 40 feet.
Form:	Broad, rounded crown. Low-level branches have secondary roots descending to around below.
Texture:	Coarse.
Leaf:	Alternate, simple, thick-leathery, dark green and shining above, paler below, 1 to 4 inches long. Pointed at both ends. Oblong.
Flower:	
Fruit:	Continuous. Inside of spheroid receptacle, fleshy, red or yellow about ¾ inch in diameter.
Geographic Location:	Native in hammocks Gulf Coast, South of Sarasota, also Atlantic Coast, South of Martin County.
Dormant:	Evergreen. Leaf-drop at 30°F. Die-back at 25°F.
Culture:	Average sandy hammock soil. Sunny, well-drained, light on moisture. Rapid growth. Somewhat subject to limb damage in hurricanes. Pests: caterpillars and scale. Quick change of foliage in March.
Use:	Valuable shade tree when existing on site, but is usually replaced on empty sites with other shade trees not as damaging to sewer lines, septic systems and water supply lines.
Other:	F. brevifolia: a compact, small tree with smaller leaves. Native.

Ficus benghalensis

BANYAN TREE

Ficus: Latin for fig

benghalensis: referring to
   Indian traders called
   "Banyans" in Bengal

Size:	85 feet x 85 feet. Original trees in India have spread by aerial roots from horizontal branches up to 2,000 feet.
Form:	Rounded haystack with great-sized trunk and many secondary aerial roots, horizontal branches.
Texture:	Coarse
Leaf:	Evergreen, broad, oval to 10 inches long, shining above and downy beneath when mature.
Flower:	Minute
Fruit:	Round, ½ inch diameter, red, borne in pairs, fleshy, fig-like seeds within the fruit.
Geographic Location:	Origin India. Introduced into islands of the Pacific and South Florida.
Dormant:	Evergreen
Culture:	Any good soil, ample water, good drainage, sunny. Fast growing. Ubiquitous root system. Control rapid spreading by removing aerial roots. Damages at 25°F.
Use:	This great tree, along with its counterpart, F. altissima, should be reserved for large, open space projects and not most residences. Often marauding roots can be slowed in groth by cutting back with a chain saw. Always a problem to underground utilities, sidewalks and asphalt driveways.
Other:	F. altissima: almost a duplicate. Differs in veination.

Ficus benjamina

WEEPING FIG

Ficus: Latin for fig

benjamina: refers to Sanskrit
name for Banyan

Size:	50 feet x 30 feet
Form:	Rounded crown with weeping habit. Aggressive root system. Usually with aerial roots in wet soil.
Texture:	Medium
Leaf:	Evergreen, pointed, up to 5 inches long, glossy.
Flower:	Minute and inconspicuous.
Fruit:	Round, red, ⅓ inch in pairs.
Geographic Location:	India and Malaya. South Florida from Palm Beach County southward. Very freeze-tender.
Dormant:	Evergreen
Culture:	Muck, marl or sandy loam. Sunny. Moist soil conditions. Can stand flooding for a limited period. Stands severe pruning with easy recovery. No pests.
Use:	Somewhat more adaptable to residential scale than the two previous ficus. Substitutes for weeping willows around ponds and lakes in Florida South. May be close-planted as a massive, soft hedge.
Other:	F. 'Exotica' is a tight-leafed, compact variety.

Ficus retusa var. nitida

CUBAN LAUREL,
    INDIAN LAUREL

Ficus: Latin for fig

retusa: notched at tip
    of rounded leaf

nitida: Latin for shining
    or polished

Size:	90 feet x 45 feet
Form:	More upright and stiffer than F. benjamina, also with a flatter crown in old trees.
Texture:	Medium
Leaf:	3 to 4 inches long, oval, blunt and short-stemmed, glossy surface.
Flower:	Minute
Fruit:	Dark red, sometimes yellow, about ⅓ inch in diameter.
Geographic Location:	India, Malaya. Now very popular in South Florida. Can resist cold damage at 28°F for a short time.
Dormant:	Evergreen
Culture:	Muck, marl or sandy loam. Sunny, well-drained. Thrives on fertilizer in early years. Fast-growing. Easily shaped by shearing. Pests: thrips.
Use:	Is prominently featured as hedges up to 10 feet high. Has been sheared into pyramids, cones and globes. Used as street trees for wide boulevards. Otherwise a handsome shade tree on large properties and park-like open spaces.

Ficus rubiginosa

RUSTY FIG

Ficus: Latin for fig

rubiginosa: Latin for rusty-red

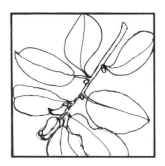

Size:	To 40 feet x 40 feet
Form:	Broad, spreading, often irregular in profile. Compact, dense foliage.
Texture:	Medium
Leaf:	Blunt-tipped, oval, 3-4 inches long. Top of leaf smooth, underside brown, hairy.
Flower:	Inconspicuous
Fruit:	Globose, 1/8 inch diameter, in pairs and warty.
Geographic Location:	Australia. In Florida: Cape Canaveral, coastal southward to Palm Beach, all of South Florida, Sarasota South.
Dormant:	Evergreen
Culture:	Almost all soils. Sunny exposure. Withstands dry situations once established. Good surface drainage. Can stand 30°F for short time. No serious pests. Salt-resistant in zones II and III.
Use:	In this genus of giants, F. rubiginosa is more adaptable for smaller areas, is a moderate grower and with some "limbing up" develops into a handsome specimen.

Filicium decipeins

FERN TREE

Filicium: Latin for
     fern-like foliage

decipiens: Latin used as
     "deceiving", referring to
     "fern" leaves

Size:	35 feet x 35 feet, often 2 or 3 trunks.
Form:	Round-headed, short trunk or trunks.
Texture:	Fine to medium
Leaf:	Compound 12-16 leaflets paired, 4-6 inches long each, winged between leaflets. Very handsome.
Flower:	Inconspicuous, small 5-parted axillary in panicles, bears both male and female flowers.
Fruit:	Purple, olive-like in clusters, small about ½ inch long.
Geographic Location:	India. Tropical South Florida only, also Hawaii and the South Pacific Islands.
Dormant:	Evergreen
Culture:	Grows easily in any soil. Slow-growing. Sun or shade. Average moisture. Good surface drainage. Fertilize twice in Summer. No known pests.
Use:	This new introduction is renowned for its fern-like foliage and thus is a fine accent plant. In Hawaii and India it is planted on roads and boulevards. Also is a handsome tubbed specimen. The fruit can be somewhat messy. We know little as to cold-hardiness, but it seems to resist any damage in the low 30°F range.

Grevillea robusta

AUSTRALIAN SILK OAK

Grevillea: for patron of botany
in England, C. Greville

robusta: Latin as in robust

Size:	75 feet high x 30 feet spread. Higher in Hawaii.
Form:	Rather densely foliated, slightly tapered broad cylinder.
Texture:	Fern-like of medium to fine texture.
Leaf:	Lacy, twice pinnate compound to 8 inches long, silky hairy on underside, new foliage appears after old leaves drop.
Flower:	One-sided clusters comb-like in appearance, 3 to 4 inches long, rich orange-fringed blossoms. Spring.
Fruit:	Black, leathery seed capsules. Year 'round.
Geographic Location:	Australia and South Africa. Central and South Florida.
Dormant:	Evergreen with intermittent leaf-drop
Culture:	Withstands 20 to 24°F when mature. Grows in poor, compact soils if not over-watered. Sandy soils in Florida. Flowers best in full sun. Stands considerable dryness. Pests: caterpillars, mushroom root rot.
Use:	Used as acacias as a nurse-crop tree, quick-growing. Has been used as a street tree and for an occasional tall hedge. Excellent reforestation subject. Not proper scale for medium and small houses. Considered fairly messy because of leaf-drop.

Guaiacum sanctum

LIGNUM VITAE

Guaiacum: Carib word for the
    gum from this wood

sanctum: Sacred or holy

Size:	To 30 feet.
Form:	Mature trees are diffuse and usually irregular. The designer's description: "with character"
Texture:	Medium to fine
Leaf:	2½ to 4 inches long. Compound, opposite, 3 to 5 pairs of leaflets broadly elliptic, 1 inch long.
Flower:	At branch tips clusters of felty blue flowers, spoon-shaped petals, twisted at base, 1 inch in diameter. Stunning in flower.
Fruit:	Obovoid, ½-¾ inch long, orange skin, edge-angled, fleshy. Black seed. Seed should be filed for germination.
Geographic Location:	Central and South America, West Indies and Mexico. In Florida sub-tropic coastal areas, Keys.
Dormant:	Evergreen, very heavy, dense wood, and will sink in water.
Culture:	Tolerates dry season conditions for months, but responds well to moisture, sun or partial shade. Tolerant to zone III salt conditions. Slow-growing. Pests: Florida red scale.
Use:	Spring blooming period is the high point in lignum vitae's annual show. Otherwise this small tree is a collector's item. It can be used as an accent shrub for many years before it grows to tree height. Native tree nurseries in Dade County are selling lignum vitae in containers now.

71

Hibiscus tileaceus

MAHOE

Hibiscus: Ancient Greek and
  Latin name

tileaceus: Latin for Tilia
  (Basswood); leaves
  are similar

Size:	35 feet x 35 feet
Form:	Usually shrubby in appearance. Branches occur from ground up. Strongly horizontal branching. Broad haystack.
Texture:	Coarse
Leaf:	Heart-shaped, dark green, nearly round, abruptly pointed, to 10 inches long, white underside, furry.
Flower:	Hibiscus-like 2-3 inches long, open as cups of clear yellow in color which changes to maroon after noon.
Fruit:	Ovoid downy capsule, 1 inch long.
Geographic Location:	Origin probably Old World. Found throughout the tropics. Sensitive to cold at 32°F.
Dormant:	Evergreen
Culture:	Tolerant of soils, sunny locations, rapid growth. Withstands some wet soil for a limited time. Subject to heeling over to its side in high winds, however recovery is rapid.
Use:	Although H. tiliaceus might be classed as a "weedy" tree because of its rapid growth and it often outgrows its location, it has its place: as high dense screen it fills its job. As a tree to frame a view or as a repeating, open-spaced tree on turnpikes and on roads in rural areas. It has been trained over a trellis on Waikiki Beach, Hawaii successfully.

Ilex cassine

DAHOON HOLLY

Ilex: Latin name Holly Oak
(Quercus ilex)

cassine: American
Indian name

Size:	40 x 18 feet.
Form:	Erect, narrow with dense foliage. Branches point upward until old, mature then irregular in shape.
Texture:	Fine to medium.
Leaf:	Alternate, leathery, smooth, elliptical, 3-4 inches long, dark green shiny above, pale pubescent below.
Flower:	White, small. Inconspicuous. Male and female on separate trees.
Fruit:	Drupe, red to almost yellow, 3/8 inch wide, Winter. Seeds 4 to 8 imbedded in flesh, pale brown, bony.
Geographic Location:	Native from Virginia to the Florida Keys and around Florida to Louisiana.
Dormant:	Somewhat dormant in northern limits. Evergreen in South Florida.
Culture:	Favors wet marginal land, swamps and stream bank. Can tolerate dryer conditions with some watering. Sandy soils or muck. Slow growth. Pests: Scale and spittle bugs. Transplants well.
Use:	A colorful specimen (female trees) when in fruit in mixed woodland-type plantings. An arboreal solution in potholes and swamps. Consider this tree as a small scale tree for residential planting. Collecting is restricted. Florida protects it.
Other:	I. opaca, American Holly: a few teeth of leaf margins, to 6 feet tall. Native in Central Florida.

Jacaranda acutifolia

JACARANDA

Jacaranda: name of Portugese-
  Brazil usage

acutifolia: sharp-leaved

Size:	60 feet x 50 feet
Form:	Upright then spreading, irregular, loose. Feathery.
Texture:	Fine.
Leaf:	Compound, opposite, finely once or twice divided feather-fashioned, having 10 to 16 pairs of divisions, each bearing 14-24 pairs of oblong pointed hairy leaflets ¼ inch long.
Flower:	Bluish-purple to mauve, bell-shaped, ½ inch wide in terminal or axillary panicles. Striking.
Fruit:	Brown circular pods 2 inches wide, splits open.
Geographic Location:	Brazil's Amazon Basin. In Florida, south and central portions.
Dormant:	Deciduous. Flowers in early Spring followed by foliage.
Culture:	No soil preference. Full sun is best. Although tolerant of dry conditions, is best when kept fairly moist. Pests: mushroom root rot.
Use:	Jacaranda's large scale pre-empts its use on most residential work. Because of Winter bareness this tree should be blended with trees which are evergreen in parks and large open spaces. It does not like windy, open sites.

Juniperus silicicola

PENCIL CEDAR, SOUTHERN
RED CEDAR

Juniperus: old classical name.

silicicola: Latin name applied
to capsules and follicles

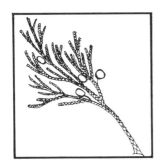

Size:	To 30 feet x 30 feet.
Form:	Early adulthood conifer-shaped, with age and space it will spread as indicated by size.
Texture:	Fine, compact and dense
Leaf:	2 typical Juniper shapes
Flower:	Inconspicuous
Fruit:	Dark blue, globose drupe with bloom. Less than ¼ inch long, smooth and resinous.
Geographic Location:	North Florida as far South as Martin County in the East and Sarasota County on the Gulf.
Dormant:	Evergreen
Culture:	Before being heavily timbered, dense stands occured at Cedar Key and other points. Sandy soils over limestone stratas. Sun or dappled shade. No demand on fertilizer other than what occurs in native soils. Pests: Juniper blight and mites.
Use:	In early days it was most popular for wind breaks planted in double lines, staggered. Most effective as in nature, in groups of random sizes with older trees in the center. Works well in open spaces on turnpikes, interchanges and rest areas. Is resistant to salt, winds and will adapt to zone II conditions if planted in Spring and allowed to adjust over Summer.

Koelreuteria formosana

GOLDEN RAIN TREE

Koelreuteria: for Professor J.
Koelreuter, Germany

formosana: for the origin
in Formosa

Size:	30 feet x 20 feet.
Form:	Broad head, low-branching, loose but attractive.
Texture:	Fine to medium
Leaf:	Compound, bipinnate, seven to fifteen 2-3 inch long leaflets, leaf to 2 feet long, medium green, soft.
Flower:	Yellow, ¼ inch long in panicales, 18 inches long. Short flowering period but very showy, early Fall.
Fruit:	Capsule 2 inches long, papery, balloon-like, bright pink to dark red, handsome.
Geographic Location:	Formosa. In Florida, cold-hardy.
Dormant:	Deciduous
Culture:	Soil-tolerant, full sun, cold-hardy. Fertilize two times per Summer. Pests: scales and mushroom root rot.
Use:	It is notable in late Summer and Fall for its production of attractive flowers and red seed capsules. Use mixed with evergreen trees to mediate its Winter look. Often in use for residential work.

Krugiodendrum ferreum

BLACK IRON WOOD

Krugiodendrum: After a tree
named for Krug

ferreum: Latin for iron, refers
to the dense wood

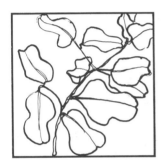

Size:	15 to 30 feet.
Form:	Crown narrow, generally regular, dense, branches erect.
Texture:	Fine
Leaf:	1 to 2¼ inch long, ovate to oval, smooth, bright to dark green and shiny, small notch in leaf tip.
Flower:	¼ inch in diameter in small short-stalked clusters, dark green.
Fruit:	Less than ½ inch, drupe, subglobose, black.
Geographic Location:	Native South Florida Peninsula, coastal areas. Hammocks, East Coast only.
Dormant:	Evergreen, sensitive to cold at 31°F.
Culture:	Densest wood in North America. Sun or shade, slow-growing, sandy soil, well-drained, heavy humus. Pests: none.
Use:	A patrician among small trees with a wonderful future. Not commonly grown in native nurseries at present. Totally resistant to high winds. Will be a fine subject for streetscape use both informally as well as regularly spaced.

Lagerstromia speciosa

QUEENS CREPE MYRTLE

Lagerstromia: for Sweden's
  botanist
  M. von Lagertroem

speciosa: splendid, showy

Size:	60 feet x 40 feet, usually less.
Form:	Broad, rounded top, not attractive branchwork when bare.
Texture:	Coarse
Leaf:	Oblong to ovate, to 12 inches long. Rough textured surface. Foliage turns red before defoliation occurs in cold weather.
Flower:	Bright pink to lavender, in pyramidal clusters. Many frilled, tubular, scentless, 3 inches in diameter
Fruit:	20-30 round woody capsules which split open to drop seed when mature, 1 inch in diameter.
Geographic Location:	India and East Indies. Tropical South Florida only.
Dormant:	Deciduous with first cold weather.
Culture:	Sunny, no soil preferences, but needs good drainage, fertilize twice in growing period; will not stand frost. Pests: scale insects.
Use:	A stunning flowering tree in bloom deserving consideration for open space projects, but blended with evergreen material to soften the unattractive winter bareness.

Lysiloma sabicu

CUBAN TAMARIND

Lysiloma: Greek for loosening
of borders; refers to
seed pod

sabicu: West Indian name
for tamarind

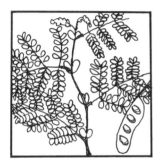

Size:	30 feet x 25 feet.
Form:	Slender tree with long, straight branches. Umbrella-like, slightly weeping. Spreading with age.
Texture:	Fine
Leaf:	Feathery, twice pinnate, leaflets oblong, ¾ inch long, light green. New flush of leaves with striking reddish tips.
Flower:	Small; white in fuzzy, globose heads.
Fruit:	Seed pod flat, thin, to 4 inches long with brown seeds.
Geographic Location:	Cuba, Greater Antilles, Trinidad and subtropical South Florida. Forest tree.
Dormant:	Semi-deciduous, subject to freeze damage.
Culture:	Prefers a rich forest loam or marl-type soil. Sunny to half-shade exposure. Good drainage and twice fertilized in the growing season. Medium growth. Drought-resistant.
Use:	A dominant subject for highway plantings from Miami to Homestead. Ideal for park planting because of its eye-catching new foliage and shade-tree character. Needs to be used more.
Other:	L. bahamensis: 60 feet x 45 feet. Umbrella profile, medium-textured. Native in Keys and mainland hammocks.

Magnolia grandiflora

SOUTHERN MAGNOLIA

Magnolia: French botanist,
P. Magnol

grandiflora: Latin for
big-flowered

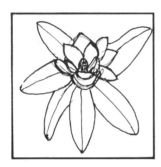

Size:	60 feet x 80 feet.
Form:	Usually a broad pyramidal and symmetrical shape until an aged forest tree.
Texture:	Coarse
Leaf:	Alternate, 5-8 inches long, lustrous, dark green surface, smooth and brown, hairy beneath, oval.
Flower:	Conspicuous, fragrant, white, 7-8 inches in diameter, heady fragrance, bowl-shaped at branch tips.
Fruit:	Cylindrical heads containing red seeds, hanging on pendant threads.
Geographic Location:	South-eastern U.S. into Florida
Dormant:	Evergreen
Culture:	Sunny or dappled shade, medium fertility as in forest loam, prefers an acid Ph, welcomes moisture, moderate to rapid growth. Tolerates some shaping. Pests: leaf spot and sun scald.
Use:	Always attracting attention as a specimen. Also can become ideal as a highway tree in natural groups of varied sizes. For residential use do not use on small properties. In California there is a great following as an espalier subject. Since there is constant leaf drop, allow low branches to cover the ground.

Magnolia virginiana

SWEET BAY

Magnolia: for early French
   botanist, P. Magnol

virginiana: for Virginia

Size:	To 30 feet in Florida. To 90 feet in northern states where also native. Often shrubby in South.
Form:	Crown irregularly rounded. Juvenile form is upright pyramidal.
Texture:	Medium
Leaf:	4 to 6 inches long, oblong to elliptic. Foliage bright green above, silvery below. Alternate.
Flower:	White, very fragrant, cup-shaped, 2-3 inches in diameter, appearing after new leaves.
Fruit:	Cone-like aggregate, 2 inches long and 1 inch in diameter. Seed flattened oval, suspended by a thread.
Geographic Location:	Native from Massachusetts, coastal plain to South Florida to Homestead. Bay heads and swamps.
Dormant:	None. Evergreen in South Florida; new leaf buds push old leaves off.
Culture:	Tolerates wet conditions on river banks and swamps. High fertility. Sunny to dappled shade. Semi-wet drainage. Tolerates pruning. Pests: scales.
Use:	Light and airy effect. Use against architectural backgrounds or evergreen screens. Birds are attracted by the red fleshy seeds.

Mangifera indica

MANGO

Mangifera: Mango-bearing

indica: of India

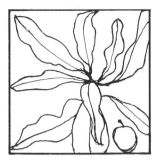

Size:	90 feet x 100 feet spread
Form:	Big umbrella-shape with stout trunk and stiff branches carrying many lateral shoots and twigs.
Texture:	Coarse
Leaf:	Narrowly oval, alternate, simple 4 to 8 inches long with slender tapering, smooth, glossy dark green.
Flower:	Orange-pink, profuse in large panicles. Fine-textured. Late Winter to early Spring.
Fruit:	Orange, crimson to green. Round, egg-shaped, usually from ½ to 6 pounds, fleshy with flat seed. Fruit matures in July and August.
Geographic Location:	India and tropical Asia
Dormant:	Evergreen. Frost-tender even in South Florida.
Culture:	Fast-growing. Needs watering and rich, black soil. Sunny location, good drainage. Fertilize three times in the growing season. Pests: Mediterranean fruit fly, subject to fungus on flowers, serious.
Use:	Other than the banana, it is the most famous fruit tree in the tropics. A street tree in Brazil for its deep shade. A drawback is the droppings of fruit on sidewalks and streets. Certain few persons are allergic to fruits and contact with the leaves.

Manilkara roxburghiana

BULLETWOOD
(MANILKARA)

Manilkara: Unknown source
in Central America

roxburghiana: William
Roxburgh, 18th
century explorer

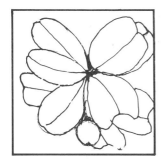

Size:	30 feet x 20 feet.
Form:	Juvenile form, has strong verticality. With age the crown develops a round head. Sometimes multi-trunked.
Texture:	Coarse.
Leaf:	Shiny dark green, 8 inches long, oblong with rounded tip with a notched apex.
Flower:	Small, white, inconspicuous.
Fruit:	Smooth, round, yellow, 1½ inch diameter. Edible. Near branch ends a single fruit issues from each leaf axillary.
Geographic Location:	Unknown. Florida Keys and South Florida from Martin County South in coastal areas.
Dormant:	Evergreen. Leaf damage at 30°F.
Culture:	Salt-resistant. Moderate to fast growth. Sunny. Tolerant to soils. Good drainage. Sensitive to cold. No pests.
Use:	One of a selected 5 trees for use in zones I and II. Its umbrella-like form lends itself to streetscapes and rather formal use with a balanced, symmetrical effect.
Other:	M. caffra: round-headed, small. Gray-green foliage, salt-tolerant.

Mastichodendron foetidissima

MASTIC TREE,
JUNGLE PLUM

Mastichodendron: meaning
Mastic tree

foetidissima: Latin for
very ill-smelling

Size:	50 feet x 40 feet.
Form:	Broad, irregular crown, stout branches, large, straight trunk up to 6 or more feet in circumference.
Texture:	Medium
Leaf:	Alternate, elliptic or oval, 3 to 8 inches long, wavy-edged.
Flower:	Pale yellow, in small axillary clusters. Gives off an unpleasant odor. Glossy above, yellow-green below.
Fruit:	Berry, globose, ½ to ¾ inch long. Juicy and yellow at maturity. Edible, acidy, somewhat bitter.
Geographic Location:	West Indies, Bahamas, South Florida from Martin County southward, also Southwest Florida coastal hammocks.
Dormant:	Evergreen
Culture:	Sandy hammock loam. Establish newly-moved trees with water, then forget them. This forest tree rises from the shady hammock floor until it reaches sunlight where it spreads its mature crown. Pests: indenture scale is common, but not fatal.
Use:	Every effort should be made to retain and preserve existing trees on all projects. Squirrels and large berry-eating birds have distributed this tree. Not a rapid-grower. A few native-tree nurseries are growing it now. Considered, along with mahogany, one of the giants of the hammocks.

Mimusops elengi

SPANISH CHERRY

Definitions obscure

Size:	Rather upright to 30 feet or more. Neat and uniform shape.
Form:	Spreading top with age. Slender stiff branching, dense foliage.
Texture:	Medium
Leaf:	Shiny, dark green, oval, 2½-3 inches long. Obtuse to bluntly acute.
Flower:	White, fragrant, about ½ inch in diameter. Corola is eight-parted with each petal two-lobed. Not showy.
Fruit:	Ovoid, 1 inch long, green to orange, 1 seed with yellow pulp. Too astringent to eat.
Geographic Location:	Malasia. Coastal South Florida
Dormant:	Evergreen. Expect some damage below 32°F.
Culture:	Sunny exposure, tolerant to dry or moist conditions. Wide range of soils. Rather rapid growth. Good surface drainage.
Use:	An excellent street tree for uniform spacing, usable 10-foot wide spaces between curb and walks. Non-aggressive root system, dooryards.

Myrica cerifera

SOUTHERN WAX MYRTLE

Myrica: Greek myric, ancient
name for tamarisk

cerifera: Latin for
wax-bearing

Size:	20 to 30 feet x 30 feet.
Form:	Crooked, gray branches support a well-balanced crown of a rather dense foliage. Sometimes picturesque.
Texture:	Fine to medium.
Leaf:	Aromatic, 3 inches long, lanceolate to oblong lanceolate. Coarsely dentate toward their tips, olive-green.
Flower:	Inconspicuous in clusters.
Fruit:	1/8 inch spheroid berries in clusters, each covered with a grayish, waxy covering. On female plants only.
Geographic Location:	Native. In moist fields and pastures, along streams and ditches. All of Florida. Cold-tolerant.
Dormant:	Evergreen
Culture:	Wet to moist conditions. Fairly sensitive to poor drainage. Sun or partial shade. Prefers sandy soils. Dying branches are no cause for alarm, this is typical of the species. Pests: caterpillars and some cankers.
Use:	Can be used for loose, informal hedges and screens. Large tree-like specimens are twigged out to form shade trees with an antique look. Foliage olive-green is good contrast when used with podocarpus, cedars and junipers.
Other:	M. cerifera pumila: newly recognized, but has great potential as a facer or ground cover.

Noronhia emarginata

MADAGASCAR OLIVE

Noronhia: Old Portuguese
name of navigator

emarginata: Latin for
shallowly notched (at
leaf tip)

Size:	Small, open-topped tree, upright form with stiff slender branching. 25' x 18'.
Form:	Upright, wide cylinder.
Texture:	Medium
Leaf:	6 x 3 inches in pairs set at a stiff angle to twigs and branchlets, grayish-green, leaf margins are curled convexly.
Flower:	Yellow, fragrant, clustered in leaf axils.
Fruit:	Purplish-green, olive-shaped to globose, 1 inch in diameter.
Geographic Location:	Madagascar and Mauritius. It grows near the shore. South Florida along coastal areas.
Dormant:	Evergreen
Culture:	Salt-tolerance in zone II with protection or zone III unprotected. No special soil requirements. Sunny exposure. Fertilize twice in growing season. No pests.
Use:	As an informal tree it is attractive in groups of 3, staggered in height planted in a triangle. It can become a good screen subject planted in a row on 10-foot centers. Individual or single trees for smallish homes soon become compact, vertical specimens.

Parkinsonia aculeata

JERUSALEM THORN

Parkinsonia: for English
    apothocary, J. Parkinson

aculeata: Latin for spine-like

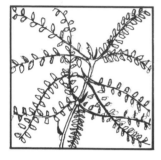

Size:	25 feet x 18 feet.
Form:	Spreading with irregular green branches with spines. Very thin with an almost ghost-like view through.
Texture:	Fine
Leaf:	Compound, bipinnate, leaflets having flattened twig-like stalks 6 inches long. The numerous small segments become deciduous in drought or cold.
Flower:	Yellow, fragrant in loose 3 to 7 inch long clusters in Spring and again in Summer. Bloom precedes leaves.
Fruit:	Pods 5 inches long, constricted between the seeds.
Geographic Location:	Mexico, Eastern California and Western Arizona.
Dormant:	Deciduous in Winter. Tolerates frost to probably 24°F.
Culture:	Tolerant of wet or dry conditions. Does well in most soils. Requires minimum maintenance. Spines are dangerous on low limbs; limb tree up for safety. Pests: in Florida thorn bugs.
Use:	From time to time the designer needs to use a "see-through" tree; look no further. Its peak performance is its bloom period in Spring. Usually performs well near banks of ponds and streams.

Peltophorum pterocarpum

YELLOW POINCIANA

Peltophorum: supporting a
pelt or shield for a
winged seed.

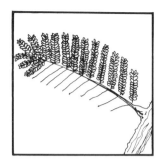

Size:	To 50 feet wide, spreading.
Form:	Much branched, handsome spreading tree with graceful, feathery foliage. Shade tree.
Texture:	Medium
Leaf:	Alternate leaves 16 x 7 inches, compound leaflets wtih opposite arrangements of ½ inch long leaflets, 20 leaflets in number.
Flower:	Buds are rusty tomentose opening to showy, yellow flowers in terminal panicles.
Fruit:	Pods to 3 inches long, flat, 4-seeded. Pods are dark red when ripe.
Geographic Location:	Brazil uplands. South and Central Florida.
Dormant:	Evergreen to semi-deciduous.
Culture:	Severe cold weather causes defoliation, but for a short time only. Fast-growing, tolerates dry conditions, best in full sun. Good drainage. Develops a very large trunk diameter.
Use:	As a flowering tree with its Spring to early Summer blanket of golden yellow, it ranks high among the flowering trees in popularity. A big-scaled shade tree for large, open areas such as in parks or institutional grounds.
Other:	P.inerme: is grown in the southern part of the state and is not as hardy. P.dubrum: cold hardy to Orlando.

Persea americana

AVOCADO

Persea: Ancient name

americana: American

Size:	To 60 feet or more, but usually not over 30 feet with an equal spread.
Form:	Upright in youth, but spreading in adulthood.
Texture:	Medium to coarse.
Leaf:	After leaf-drop come blooms. New leaves come thereafter, oval 3 to 16 x 2 to 10 inches, downy when young.
Flower:	Small, greenish-white in early Spring occuring in close, terminal panicles.
Fruit:	South American avocado is green-skinned and up to 3 lbs. Guatemalan avocado is brown-skinned and smaller.
Geographic Location:	Brazil's Amazon Valley, Central America, West Indies and in subropical Florida.
Dormant:	Evergreen
Culture:	Small, brown-skinned fruits in Mexico. 25°F is sufficient for saving trees, but early flowers will be frozen. Good drainage. Fertilize lightly. Use fallen leaves under trees as a mulch. Remember, this is a forest tree in its own jungle habitat. Pests: scales.
Use:	The avocado is a dooryard tree, fruit tree and grove tree. It is treasured because of the "buttery" meat beneath the skin. In Florida the avocado and the mango, along with the various citrus trees have their place in most residences. Selected varieties can extend the season for 6 months.

Persea borbonia

RED BAY

Persea: Ancient name

borbonia: obscure

Size:	30 to 40 feet
Form:	Slender trunk, spreading crown, dense, branches erect.
Texture:	Medium
Leaf:	Aromatic, smooth, shiny above, 2½ to 6 inches long, bright green above, paler below, elliptic, tips pointed.
Flower:	Small, greenish, on stalks, axillary in few-flowered clusters.
Fruit:	Matures in Fall, blue to nearly black, ovate, ½ inch long. Pale, brown, bony seeds.
Geographic Location:	Delaware to Florida to Texas in hammocks and swamps and coastal plains.
Dormant:	Evergreen
Culture:	Native. Sun or dappled shade, tolerates wet, swampy situations as well as dry-season sites. Sandy soils. Pests: red bay scale.
Use:	This tree substitutes for European bay leaves in cooking; almost as aromatic. This is a fine ornamental and shade tree available now in native plant nurseries. Use with other natives for developing a hammock. Occasionally is handicapped with random die-back.

Pimenta officinalis

PIMENTO, ALLSPICE,
    JAMAICA PEPPER,
    GUANGO

Pimenta: Latin-spanish
    for pepper

officinalis: used in medicine

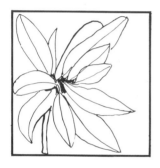

Size:	To 40 feet.
Form:	Upright, ovoid shape, dense, bark scales off in sizeable patches giving a bicolored effect.
Texture:	Medium
Leaf:	Dark green, stiff, oblong, to 7 inches long, with short stems. Aromatic leaf is used for Bay Rum.
Flower:	Small, white, clustered at branch terminal.
Fruit:	In clusters, ¼ inch in diameter, brown and are the commercial, spicy product known as allspice.
Geographic Location:	Jamaica, West Indies, South Florida coastal sub-tropics. Dry hillsides in Jamaica.
Dormant:	Evergreen. Cold-tolerant to 27°F.
Culture:	Adapts to moist conditions easily, but also does well under dry-season situations. Does well in dappled shade or sun. Tolerates many soils. Pests: unknown. Slow-growing.
Use:	This tree is a good addition to the small home, or fits into the natural mixed shrub border as an accent. Its grayish bark stands out in strong moonlight. Should be used more.

Pinus elliottii

SLASH PINE OR
   CARIBBEAN PINE

Pinus: Latin for the pine

elliottii: for early
   American botanist

Size:	To 100 feet tall.
Form:	Tall, slender, spreading at top with horizontal branches. Regular brush fires burn off lower limbs, self-pruning.
Texture:	Fine
Leaf:	Needle-like 8-12 inches long in sheaths of 2 or 3 leaves on the same tree, glossy green tufted at the ends of branches; stiff, dropping the second season.
Flower:	Inconspicuous
Fruit:	Cones from 2 to 6 inches long, ovoid, conic. Scales thin, lustrous, brown, armed with a sharp prickle.
Geographic Location:	Native to South Florida, thence coastal Florida on both Atlantic and Gulf shores to Central Florida. Also found as far South as Honduras and the Greater Antilles and Bahamas.
Dormant:	Evergreen
Culture:	Any soil, wet or dry conditions. Sunny. Gregarious in reseeding. Pests: Pine bark borer.
Use:	Almost every landscape design in native material should reintroduce colonies of this pine. Also is used in mixed woods being planted on berms designed as buffers on recent projects. Excells for reforestation of old fields.
Other:	P. elliotti var. densa: is the famous "Dade County Pine" of hardwood fame. Now almost extinct.

Piscidia piscipula

JAMAICA DOGWOOD,
   FLORIDA FISHPOISON
   TREE

Piscidia: Latin for fish poison

piscipula: Latin for small fish

Size:	To 40 feet with a trunk height of 1½-3 feet above ground line.
Form:	Ascending stout branches with an open, broad or narrow crown, irregular.
Texture:	Medium to coarse.
Leaf:	Alternate 7-9 leaflets, on stem, ¼ inch long, oval. 3-inch leaflets, smooth, firm, dark green above, paler below.
Flower:	Pea-like white and lavender, ¾ inch long in elongated stalks, 2-5 inches long, fragrant.
Fruit:	A 4-inch pod, light brown with 4 papery scalloped wings, several red-brown, flat seeds.
Geographic Location:	Native. Florida Keys, Bahamas, West Indies.
Dormant:	Evergreen, sensitive to frost. Defoliates for about 4 to 6 weeks, but flowers first.
Culture:	Sandy loam, full sun, will tolerate dry conditions. Can be grown on barrier islands. Atlantic Coast North to Vero Beach. Rapid growth.
Use:	The subject's greatest beauty is displayed when masses of misty, lavender flowers are produced between old and new leaves giving the appearance of tree-type wisteria. Highway beautification when blended with other trees. Reflective image at shore-line of ponds.

94

Pisonia discolor

BLOLLY

Pisonia: Pea-like, of
　　leaden color

discolor: of 2 colors as in upper
　　and lower sides of a leaf

Size:	30 to 40 feet high, trunk 12 to 20 inches.
Form:	In open situations a spreading crown, irregular in dense forests, short, compact, round-topped.
Texture:	Medium
Leaf:	Opposite or alternate, simple, 2 to 3 inches long, ½ to 1 inch wide, oblong, obovate; apex rounded or notched, shiny.
Flower:	Fall, in few flowered terminal or axillary clusters, no petals, greenish-yellow.
Fruit:	An enlarged red, fleshy drupe with longitudinal ribs, ½ inch long.
Geographic Location:	Southeastern subtropical Florida in Keys and on the mainland. Native in coastal hammocks. Cuba, West Indies, Bahamas.
Dormant:	Evergreen
Culture:	Sandy hammock loam. Moderate water. Full sun or dappled shade. Moderate growth. Stands temporary dry periods when established. Good drainage. No pests.
Use:	As a large shrub or small tree it is best known especially when it has produced its handsome fruits, which are borne in quantity. Leaves are variable when planted from seed.

Plumeria acuminata

FRANGIPANI PLUMERIA

Plumeria: for French botanist,
C. Plumier

acuminata: Latin, making an
acute angle, refers
to flower

Size:	20 feet x 15-20 feet.
Form:	Small, stiff, broad-crowned tree with thick branches.
Texture:	Coarse
Leaf:	Alternate, smooth, oblong, pointed to 16 inches long, dropping in late Fall, thickest at ends of branches.
Flower:	Fragrant, white and yellow, narrow, tubular, each about 2 inches in diameter. Spring and Summer, before leaves.
Fruit:	8 to 12 inches, forked, twin pods ¼ inch in diameter, brown, containing numerous flakes each carrying 9 seeds.
Geographic Location:	Tropic America. Now throughout the tropics. Coastal areas in South Florida.
Dormant:	Deciduous from November to late April.
Culture:	Moist, rich loam. Full sun. Some varieties are shrubby and should be pruned up to develop tree-like character.
Use:	The different species have different growth habits and forms. All blend well in blooms and colors. A really effective massing can be attained by mixing them. To soften the deciduous effect, mix with broad-leafed evergreens.
Other:	P. rubra: pink to red flowers, numerous hybrids. P. Singapore: hybrid, white flowers, obtuse leaves.

Podocarpus gracilior

WEEPING YEW,
FERN PODOCARPUS

Podocarpus: Greek for
foot and fruit

gracilior: Latin for slender
and graceful

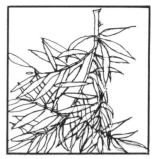

Size:	30 feet x 25 feet.
Form:	Narrow foliage is held by slender drooping branches. Rounded crown with age.
Texture:	Fine, almost fern-like.
Leaf:	Soft, closely spaced, bluish green or grayish, 1 to 4 inches long, narrow, ¼ inch.
Flower:	Inconspicuous catkin or scale, enclosed ovules.
Fruit:	Drupe-like, red aril attached, edible.
Geographic Location:	Central Africa, tropical.
Dormant:	Evergreen, sensitive to frost at 28°F.
Culture:	Sun or shade, good drainage. Not drought-tolerant. Slow growth. Ordinary topsoil. Responds to fertilizer: twice per growing season. Pest-free.
Use:	Often used in espalier-fashion on cool walls, also makes a soft hedge, entry or accent plant. Excellent streetscape subject with clean foliage. Handsome top and shade tree. Weeping habit makes it attractive on borders of ponds.

Podocarpus macrophyllus
  'Maki'

JAPANESE YEW TREE,
  YEW PINE

Podocarpus: Greek for foot
  and fruit

macrophyllus: Greek for
  large-leafed

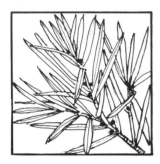

Size:	To 50 feet in Japan; 15-20 feet in Florida.
Form:	As a variety of P. macrophyllus. Its slower growth, left unsheared, develops a shrubby form.
Texture:	Fine
Leaf:	Flat needle lanceolate, 3/16 inch x 4 inches long, dark glossy green. Maintains foliage to ground.
Flower:	Catkins in male form, aril in the female, not conspicuous.
Fruit:	Drupe-like with fleshy aril attached. Bluish-plum colored.
Geographic Location:	Japan. Frost-tolerant. South and Central Florida.
Dormant:	Evergreen
Culture:	Plant in topsoil for rapid growth. Later adjusts to most Florida soils. Takes well to pruning. Sunny, but semi-shade tolerant. Fertilize twice in growing season. Pest-free.
Use:	Has been pruned and trimmed into all the old forms: coniferous, columnar, spherical, etc. But unusual, natural, irregular forms can be picturesque using multi-stemmed natural subjects.

Pongamia pinnata

PONGAM

Pongamia: obscure

pinnata: pinnate foliage
    arrangement

Size:	30 to 50 feet x 25 x 35 feet.
Form:	Broad, shallow crowned tree with large trunk.
Leaf:	Compound, 5 to 7 leaflets up to 4 inches long, pinnate.
Flower:	Pale lavender, 5-inch long pendant in clusters. Fragrant, pea-like.
Fruit:	Pod, woody, contains one seed, 1½ inch long.
Geographic Location:	Tropical Asia, Australia. South Florida coastal areas to Vero Beach on barrier islands.
Dormant:	Evergreen except for a brief change in early Spring, usually March or sometimes February.
Culture:	Muck, marl or sandy soil. Sun to partial shade, fast-growing. Good drainage. Moist conditions. No serious pests.
Use:	As a street tree dropping seed pods are messy on sidewalks otherwise excellent for the purpose. Tree makes a fine specimen for large open spaces. Resistant to high winds and long dry periods.

Prunus caroliniana

CAROLINA CHERRY
LAUREL

Prunus: classic Latin
for plum

caroliniana: Carolina, U.S.A.

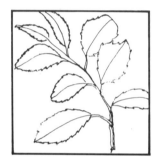

Size:	20-30 feet x 15-20 feet with age.
Form:	As a shrub upright to ovoid, as a mature tree with strong verticality then to rounded crown.
Leaf:	Medium
Flower:	Small, white, fragrant in racemes, flowers 1/8 inch in diameter, 5-petaled.
Fruit:	Small cherry-like drupes, ½ inch in diameter, purple-black enclosing 1 stone
Geographic Location:	Native in coastal hammocks in South Florida especially in the Stuart area. Also occasional in Central and North Florida.
Dormant:	Evergreen in the South.
Culture:	Growth rate rapid, sun and shade, average fertilizer, good drainage and good moisture especially up to 8 feet. No pests.
Use:	Excellent as a small background tree, as an informal screen or buffer. Grown as a specimen and noted for its dense foliage. Sheared as a hedge, shaped as in a column or pyramid, pruned into a formal standard or "lollipop".

Prunus myrtifolia

WEST INDIAN CHERRY

Prunus: classic Latin for plum

myrtifolia: Latin for
myrtle-leafed

Size:	To 45 feet.
Form:	Juvenile shape is pyramidal graduating to rounded crown with age.
Leaf:	Elliptic, often acuminate, glabrous, glossy upper surface, about 2¼ to 4½ inches long.
Flower:	In axillary racemes shorter than the leaves, white, about 1/8 inch in diameter.
Fruit:	Subglobose, cherry-like, glossy black-purplish about ⅓ inch in diameter with one stone.
Geographic Location:	Native in South Florida in hammocks, also Bahamas, Cuba and Greater Antilles. Native.
Dormant:	Evergreen
Culture:	Growth rate is rapid, often an understory subject. Shade or sun. Good drainage, average moisture. No pests.
Use:	Can be grown as a specimen, as a background loose screen or tight hedge. Little known among landscape people and is now grown by several Dade Coutny nurserymen. It will gain popularity.

Psidium littorale

CATTLEY GUAVA

Psidium: the name for pomegranate in Greek

littorale: pertaining to the beach or shoreline

Size:	25 feet x 12 feet.
Form:	A semi-upright shrub or small tree often with 2 or 3 narrow trunks, multi-branched and mottled.
Texture:	Medium
Leaf:	Opposite, thick, leathery, 4 inches long, deep green, waxy upper surface.
Flower:	White, many stamens, similar to powder-puff in miniature. 1 inch in diameter, axillary.
Fruit:	Red pome, 1½ inch diameter, sweet acid with many small seeds like grape seed, delicious.
Geographic Location:	Brazil. In Florida in subtropic coastal areas.
Dormant:	Evergreen
Culture:	Prefers rich soil, but adaptable. Drought-tolerance when established. Raccoons and squirrels as well as birds create volunteers. Fertilize twice in growing period. Sunny. No pests.
Use:	Multi-trunked selected trees, when judiciously pruned, present an almost exotic or Japanese picture as an accent. When used as a hedge keep the plants fairly loose for maximum effect. Fine dooryard tree.
Other:	P. l. lucidum: lemon guava, introduced from California, slower growing and slightly larger.

Quercus laurifolia

LAUREL OAK

Quercus: Classic Latin for oak

laurifolia: with leaves like
 the laurel

Size:	40-60 feet x equal spread or more.
Form:	Dense, upright, developing in age to large haystack crown.
Texture:	Medium
Leaf:	Alternate, entire or faintly lobed, semi-evergreen, elliptic, broadest above the middle.
Flower:	Catkins, pistilate in spikes, Spring.
Fruit:	Acorn, 5/8 inch in diameter, light brown; ovoid acorn, cup deeply saucer-shaped.
Geographic Location:	Native, Virginia to Florida and Louisiana. Sandy hammocks, river banks and wetlands.
Dormant:	Deciduous to semi-evergreen.
Culture:	Prefers moist to wet conditions, rich, sandy loam. Rapid growth. Life span 50 to 60 years. Ideal for use as a water survivor in its favorite locations, such as river plains. Sun or shade.
Use:	Combine as a large forest tree with red maple, cassine holly, carolina willow and wax myrtle. Its short life span preempts its use as a street tree.

Quercus virginiana

LIVE OAK

Quercus: Latin for oak tree

virginiana: for Virginia

Size:	To 50 feet x 30 to 50 feet spread.
Form:	Short, massive, very wide spread with horizontal branches.
Texture:	Medium
Leaf:	Alternate, ovate to oblong, 2-5 inches long x 1-2½ inches wide, shiny. Late March to April.
Flower:	Catkins clustered 2-3 inches long. March.
Fruit:	Elliptic acorn ¾ inch to 1 inch long, nearly black.
Geographic Location:	Virginia to Florida. Cuba, Mexico, Central America, also to Louisiana near the coast.
Dormant:	Deciduous to semi-evergreen.
Culture:	Prefers light sandy loam, well-drained, sunny location. Reacts well to feeding. Moderate growth. No serious pests.
Use:	Ideal native tree for shade, open spaces: parks and schools and where there are generous rights of way for boulevard or highway planting. In time, oaks will form an archway over the paving. The dappled shade of the live oak is the right intensity for ground covers like ophiopogo, liriope, aspidistra and wandering jew.

104

Schefflera actinophylla

QUEENSLAND UMBRELLA
    TREE

Schefflera: obscure

actinophylla: leaves arranged
    radially

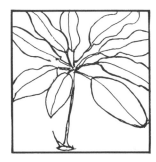

Size:	25 feet by 12 feet.
Form:	Single to multi-stemmed trunks generally upright and each carrying a smallish crown.
Texture:	Coarse.
Leaf:	Quinquefoliate, palmately compound; leaves on 1½ to 2 feet leafstalks.
Flower:	10 to 15 spikes with a series of showy yellowish flowers, 1 inch in diameter borne in clusters.
Fruit:	10 to 15 flower spikes support large ovoid berries, rust-reddish. These spikes are likened to the ribs of an umbrella.
Geographic Location:	Australia
Dormant:	Evergreen
Culture:	Fast-growing; tolerates sun or shade. Prefers rich, moist soil; must have good drainage. Seeds are spread by birds, thus now a volunteer. No serious pests.
Use:	By judicious cutting back of 1 or 2 trunks to extend limited height of the tree, this around pools. Probably over-used since World War II. Forms a buffer/wind-break and screen at property lines. A fairly clean tree with minor leaf-drop.

Simaruba glauca

PARADISE TREE

Simaruba: Carib Indian for
bitter tonic made
from tree

glauca: Latin for silvery
undersides of leaves

Size:	50 feet x 25 feet.
Form:	Upright in early years to rounded crown in adulthood. Dense foliage and many branches.
Texture:	Medium
Leaf:	Compound, 10 to 20 leaflets 2 to 4 inches long. Leaf from 6 to 10 inches. Pinnate. New leaflets reddish.
Flower:	Large panicles of lightly yellow flowers, tiny. Dioecious.
Fruit:	To ¾ inch in April, ellipsoid drupes from green to red to plum-purple. Clusters of 3 to 5.
Geographic Location:	Native in subtropic Florida coast to Cape Canaveral to Key West including Naples area on West Coast.
Dormant:	Evergreen.
Culture:	Rich, sandy soil high in forest humus. Rapid growth. Shade to sun. Good drainage. Easily moved as a young plant or seedling. Apparently no insect pests, occasionally sooty mould.
Use:	A great native for use in the warm areas to build a hammock with, including gumbo limbo, oak, satin leaf, red bay, mastic and sabal palms. Caution on use near parked cars, sidewalks, paved patio and pool deck areas. Seeds and fruit are messy and stain hard surfaces.

Swietenia Mahagoni

MAHOGANY

Swietenia: After Dutch doctor
  Baron von Swieten,
  18th century

mahogani: of unknown origin.
  Perhaps Carib Indian

Size:	To 75 feet.
Form:	Broad cylinder to oval shape.
Texture:	Medium
Leaf:	Compound with 4 to 5 leaflets which are elliptic to broadly ovate, 2 to 4 inches long.
Flower:	Inconspicuous, greenish on stalks in leaf axils.
Fruit:	Ovoid brown capsule 5 inches long, erect which, when ripe, splits from the base releasing wing seeds.
Geographic Location:	Florida's upper Keys, Cape Sable in hammocks as a native long ago logged out. Now a captive native. Stands more cold then many other natives.
Dormant:	Evergreen except for a 2 week period in Spring.
Culture:	Rich forest loam and humus. Early shade to full sun. Fast growing. Good drainage. Responds to fertility. Pests: leaf notcher, scale, webworm and leaf miner.
Use:	Symmetrical avenue plantings. Also good in natural streetscapes. A favorite shade tree for large properties as well as parks, industrial parks, etc. Semi-salt resistant.
Note:	When planting a repetitive row, all trees must shed and replace foliage in unison. This can be done only by vegetative reproduction from one clone.

Syzigium cumini

JAMBOLAN PLUM

Syzigium: Latin, refers to
   flower petals fused
   together in a cap.

cumini: obscure Latin name

Size:	120 feet x 40 feet. Gigantic in scale with age.
Form:	Always a broad cylindrical shape, sometimes a bit wider at base; short trunk, wide diameter.
Texture:	Medium to coarse.
Leaf:	Broadly oblong to ovate, 8-10 inches long, opposite, short-stemmed petiole.
Flower:	Petals white in branched cymes fused into a cap. Flowers funnelform.
Fruit:	Berry is ovoid, plum-like 1½ to 2 inches long in clusters, purplish-red, occasionally delicious and fleshy, single pit or stone.
Geographic Location:	India, East India, South Florida subtropics.
Dormant:	Evergreen
Culture:	Prefers rich black soil. Tolerant to heavy and light moisture conditions. Welcomes fertility, fast-growing. Good to medium drainage. No serious pests. Sun or light shade.
Use:	This tree excells as a wind-breaker. Use it in open spaces where shade tree habits can develop such as airports, parks, industrial parks or cemeteries. Will dwarf one-story buildings. Fruit when ripe can be messy on pooldecks, patios, and public walkways; but welcomed by birds.

Tabebuia caraiba

SILVER TRUMPET

Tabebuia: old name of
Brazilian origin

caraiba: Indian name of the
Caribbean Basin

Size:	To 25 feet.
Form:	Corky bark trunk, crooked, erect with compact head. Often develops a leaning character.
Texture:	Medium to coarse.
Leaf:	Deciduous, divided into 5 to 7 somewhat stiff, narrow leaflets to 6 inches long. Silvery on both sides, clustered at ends of branches.
Flower:	Bell-shaped up to 2½ inches long in profuse clusters before new foliage. Butter-yellow.
Fruit:	Pod to 4 inches long by 5/8 inch diameter, is grey, wafer seed in membrane.
Geographic Location:	Native to Paraguay, South America. Coastal South Florida.
Dormant:	Deciduous in Winter. Not cold-hardy.
Culture	Moderate growth rate. Prefers full sun, average moisture and good drainage. The rather weak root system results in irregular character. No pests.
Use:	A most important Spring-bloomer. Its upright habit gives it reason for use in small or restricted areas. When it combines with evergreen trees it has its purpose in streetscapes.

Tabebuia heterophylla

PINK TAB OR TRUMPET
TREE. ROBLE: SPANISH

Tabebuia: Latinized
  Brazilian name

heterophylla: having leaves of
  more than one shape

Size:	To 50 feet.
Form:	Slender pyramid. Crown broadens with age.
Texture:	Medium
Leaf:	5 palmate compound leaflets in 2 opposite pairs, leaflets of unequal size, up to 6 inches long, elliptic oblong.
Flower:	Pink or whitish slender bells to 3 inches long in clusters, irregularly toothed, occur at branch ends.
Fruit:	A slender pod up to 12 inches long containing many seeds, each imbedded in a papery membrane.
Geographic Location:	South and Central America, West Indies. In Florida, from eastern Palm Beach County South.
Dormant:	Semi-deciduous. Flowering before new foliage in Spring.
Culture:	A wide range of soils from semi-dry mountains to wet stream beds. Organic fertilizers twice per growing season. Full sun and good drainage. No pests. Subject to damage from frost.
Use:	For formal street planting use vegetative propogation from a single clone for uniform blooming. A winner used in informal tree plantings combined with broad-leafed evergreen trees. This is well adapted to residential sites.
Other:	T. bahamensis is of the pallida species, but a pretty little dwarf to 7 feet average.

Tamarindus indica

INDIAN TAMARIND

Tamarindus: from the Arabic

indica: Indian

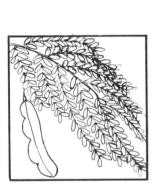

Size:	To 80 feet x 50 feet.
Form:	Open spreading crown, feathery foliage. Domed with age.
Texture:	Fine to medium.
Leaf:	Dense pinnately compound with each leaf having 10 to 20 pairs of leaflets of ½ inch length.
Flower:	Red and yellow to 1 inch wide, in small racemes with a few flowers.
Fruit:	Brittle-skinned pods 4 to 6 inches long containing a sweet-sour pulp surrounding two or three seeds.
Geographic Location:	India, East Indies, South Africa. In Florida subtropics along the coast of the Atlantic and the Gulf.
Dormant:	Evergreen. Temperatures above 32°F.
Culture:	Deep, rich soil with abundant rainfall or irrigation. Full sun. Slow rate of growth. Pests: subject to root mushroom rot.
Use:	Caravan journeyers chewed the pods to slake thirst. Very durable in high winds. The sticky pods, when dropped on hard surfaces, can be objectionable. A feathery, graceful tree for use as an open space subject.

111

Taxodium distichum

BALD CYPRESS

Taxodium: Greek for order
and arrangement

distichum: Latin for a stem
carrying opposite leaves.

Size:	To 100 feet in Florida.
Form:·	In youth, is upright, pyramidal; in age, broad-topped.
Texture:	Fine
Leaf:	2 ranked needles ½ inch long, yellow-green. Wide leaf 4 inches long. Leaves turn olive to red in Fall. Bare in Winter.
Flower:	Male panicles 4 inches long.
Fruit:	Cones ¾ to 1 inch in diameter, globular winged seeds.
Geographic Location:	Native to Texas, Florida and North in Mississippi Valley to Illinois. Stays smaller in the North.
Dormant:	Deciduous
Culture:	In fresh-water swamps flooded for at least part of the year. Prefers acid reaction to alkaline soil. Shows a remarkable range of climatic zones. No serious pests.
Use:	Although loved for its feathery summer coat of light green, its bare-bones branchwork and trunk in the Winter is also worthy of appreciation. Plant in staggered sizes in colonies at stream banks, wetlands and shorelines of lakes.

112

TALL SHRUBS

The classification represents a category of plants used in an average bracket of from six feet to twelve feet. It is possible that a few small trees could fall within this mean, however the section on trees will retain and list those subjects.

Tall shrubs have as much claim to demand and use as have the other sections on shrubs and ground covers. They meet the requirements as background plants for ground covers and medium shrubs.

Their use as screen material to hide unwanted views or for privacy buffers is high on the list. Consider the life of these plants as compared to a wood fence with a life of eight to ten years, and they need no stain or painting and withstand high storm winds that destroy fences.

Underlined should be the presentation of seasonal color if needed, such as oleanders, Powder Puffs, Yellow Elders and the bird-attracting fruit of the Pyracantha and others.

Acalypha wilkesiana 'Musaica'

COPPER LEAF

Acalypha: Latin, without a calyx

wilkesiana: For Charles Wilkes, South Pacific explorer, era of 1840.

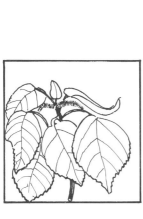

Size:	8 to 15 feet high
Form:	Upright, but with a well-rounded head. Dense, sometimes needs shaping.
Texture:	Coarse
Leaf:	4 to 8 inches long. Bronze green to muted red, mottled or mosaic-patterned, heart-shaped, semi-rolled.
Flower:	Tiny flowers are borne on slender spikes. Male spikes are pendant while female spikes are upright, both to 12 inches long.
Fruit:	Capsule, but insignificant
Geographic Location:	Fiji Islands, cannot stand near-freezing temperatures.
Dormant:	Withstands drought and poor soils; luxuriant.
Culture:	Of rapid growth. Adjusts to many soils, prefers sunny exposures, requires moisture. No major pest problems. Easily propagated by air-layers or mist-grown cuttings.
Use:	An old reliable of the borders and edgers in Victorian gardens and public parks. Its use today might well be relegated to Disney or Fairgrounds as dynamic accents or hedges.
Other:	A. godseffiana. Superior, but conservative: heart-shaped, dark green with a margin of creamy white with a toothed edge.

Ardisia escallonioides

MARLBERRY

Ardisia: Arrow shape of the
flower's corolla lobes.

escallonioides: Refers sim-
ilarly to escallonia.

Size:	10 to 15 feet — usually much taller than wide.
Form:	Semi-upright
Texture:	Medium
Leaf:	3 to 4½ inches long by 1½ inches, dullish semi-glossy surface, dark green. Well defined mid-rib, oblong and pointed leaf.
Flower:	Small, white in dense terminal panicles, fragrant, of short duration.
Fruit:	Many, purple, ¼ inch diameter in late Spring.
Geographic Location:	Native, occurs as understory below Sabal Palm clusters, on dry ground and often occuring with native stoppers.
Dormant:	Evergreen
Culture:	Prefers well-drained semi-shade to full sun, in soils of sandy humus to calcareus shell soils in pine lands and hammocks. Fast growing, no serious pests. Was nursery grown before World War II.
Use:	A must-use plant in native planting design in mixed groups as background for psychotria and hamelia, etc.

Bauhinia galpinii

NASTURTIUM BAUHINIA

Bauhinia: For the brothers
   Bauhin, European
   herbalists

galpinii: For Ernest Edward
   Galpin, 1858-1941

Size:	10 feet x 10 feet
Form:	Dense, haystack shape. Many descending exterior branch tips. Semi-climber. Scattered groups of flowers.
Texture:	Medium
Leaf:	Alternate, 2 inches x 2½ inches wide, rounded with a deeper than ¼ inch cleft mid-way at top pf leaf, 4 radial veins, light green.
Flower:	Five petals, 1½ inches wide, each one shaped like the ginko leaf with 3 stamens reaching upright 1½ inches long. 3 to 5 flowers on a stalk, axillary.
Fruit:	Dark brown pods to 5 inches long in late Summer containing 3 to 7 shiny seeds, semi-flat.
Geographic Location:	Origin, tropical Africa, but now well distributed thoughout the warm belts of the world. Tropic only.
Dormant:	Semi-defoliated during Winter.
Culture:	Well-drained conditions, warmth, sunny exposure only. Sandy loam-type soils. Responds quickly to fertilizer. Needs early Spring shaping or pruning for growth control.
Use:	Multi-use bougainvillea-like mound. Excellent trellis-type climber, espallier, large planter boxes, good lawn specimen in linear evenly spaced design. Large tubs and containers.

Caesalpinia pulcherrima

BARBADOS FLOWER-FENCE

Caesalpinia: Tufted warm-
loving plants

pulcherrima: Most beautiful,
as in a girl

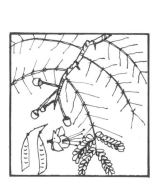

Size:	10 feet x 10 feet
Form:	Open, loosely branched and gracefully upright
Texture:	Fine
Leaf:	Feathery, compound then doubly compound feathery leaflet, forked with leaflets ½ inch roundly ovate. Color is light green.
Flower:	Warm red and butter yellow, showy five petals in clusters. Stunning flower color all Summer.
Fruit:	Pods turning dark brown 4 inches long, ¾ inch wide. Persistent, easy germination of seed.
Geographic Location:	Native to the W. Indies. Now distributed thoughout the world tropics.
Dormant:	Evergreen except when frozen, but recovers from the roots in warm weather.
Culture:	Any soil as long as it is well-drained. Best grown in full sun. Light waterings to dry conditions after being established. Pests; nematodes and scales.
Use:	Valued for its long duration of flowering. Good backdrop as a semi-screen. Does not mix or blend well, dense, coarse-textured tropicals. Better used with feathery or fern-leafed plants of fine or medium texture.
Other:	Variety Flava is the all yellow clone, which mixes well.

lliandra haematocephala

POWDER PUFF

Calliandra: Calli refers to
  beautiful, andra refers to
  stamens

haematocephala : A blood-red
  head, as in the flower.

Size:	10 or more feet x equal spread
Form:	Large dense handsome round shrub
Texture:	Medium to fine
Leaf:	Alternate dark green leaves made up of 2 forked leaflets in 6 even pairs. Each leaflet is 1¾ inches x ½ inch in an unequal ellipse.
Flower:	Large blood-red powder puffs, 3 inches across of silky stamens in late Fall and Winter in South Florida.
Fruit:	Straight brown pods with thick margins.
Geographic Location:	Bolivia
Dormant:	Evergreen in Central and South Florida except for killing freezes. Some recovery from the roots.
Culture:	Rapid growth, sandy soils in full sun. Not demanding of artificial waterings. Occasional shearing to maintain good form may be needed. Pests: Occasional mites and worms.
Use:	Ten foot plants on eight foot centers forms a non-penetrating hedge of estate quality. Individual specimens are beautiful in flower in an open lawn. Should be part of your pallet.

Cassia beariana

Cassia: Old Greek name for the senna plant of early medical use

beariana: After the man's name

NOTE: C. beariana is a subspecies of C. abbreviata

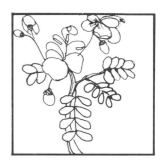

Size:	To 10 - 14 feet by 8 - 10 feet wide
Form:	Haystack-shaped low branching
Texture:	Fine
Leaf:	Usually 4 pairs of leaflets 2 inches long by 1 inch wide each compound leaf is about 4 inches long. Alternate light green, evergreen.
Flower:	Cluster of rich butter-yellow occuring roughly at Thanksgiving and again at Easter.
Fruit:	Pods 6 to 8 inches, cylindrical, brown when mature.
Geographic Location:	S.W. Africa, South and Central Florida and fromerly as far North as Glen St. Marys and Jacksonville.
Culture:	Very simple requirements as to soil and watering. Does best in full all day sun, good drainage with occasional feeding. For best appearance as a round, compact shrub some cutting back is in order.
Use:	Beareana's two season blooming periods, when it knows no peer, makes it ever popular and demanded during those periods. A slight drawback is it is not long lived, over 5 - 10 years.

ım nocturnum

NIGHT BLOOMING JASMINE

Cestrum: A plant of Greek
origin

nocturnum: Belonging to
the night

Size:	8 to 12 feet with 6 to 8 feet spread
Form:	Rather loose and soft in appearance, semi-erect
Texture:	Medium
Leaf:	Leaves 4 to 8 inches long. Shiny oval to oblong on slender branches.
Flower:	Chartreuse to greenish-white, about ¾ inch long, the short lobes pointed and erect or spreading, occuring in Summer, fragrant at night.
Fruit:	One third inch in diameter, white and juicy in fair quantities. Said to be toxic, one or two seeds.
Geographic Location:	West Indies, in Florida's southern regions. Coastal areas.
Dormant:	Evergreen
Culture:	Sunny locations with plenty of moisture and fertilizer. Should be clipped back after fruiting to maintain decent form.
Use:	Because of the super-heady perfume exuded at night, caution is advised concerning proximity to residences when locating this plant. Best advice is to position the plant a distance away where the night breezes will drift the essence toward a porch. Blooming occurs during the full moon only.

Chiococca alba

SNOWBERRY

Chiococca: Greek for
    Snowberry

alba: The color white
    describes snowy fruit

Size:	From 2 feet to 10 feet
Form:	The low scrambling shrub is quixotic when offered a little support as it climbs as high as 10 feet.
Texture:	Medium
Leaf:	Shiny dark green elliptical to 2¾ inches long, ¾ inches wide. Can be acute-tipped to obtuse. Opposite flowers branching laterally from both axis of the opposite leaves in scapes.
Flower:	Born in axillary racemes or panicles, tipped with 6-8 tubular flowers 7-8 mm. long. White turning yellow.
Fruit:	Ovoid to orbicular 5-7 mm. wide. Sparkling white.
Geographic Location:	Native hammocks and pine woods
Dormant:	Evergreen
Culture:	Most native, sandy soils including the loams of hammocks. Sunny to dappled shade. Tolerates dry conditions with flushes of growth in late Spring. Survives without care. Pests: chewing insects, occasionally.
Use:	Chiococca can be collected easily as young plants and adjusts easily to new locations. Performs well as a trellis or espaliered subject. Can be used as a mounding specimen.

Chrysobalanus icaco

COCO PLUM

Chrysobalanus: Greek for
"golden acorn"

icaco: Arawak term
for cocoplum

Size:	To 30 feet, predominantly a 10-12 foot x 8 foot shrub.
Form:	Usually a symmetrical compact "dome" fully leafed from the ground up.
Texture:	Medium
Leaf:	Alternate rounded in 2 rows turned upward along the twig, glossy, leathery, up to 3½ inches long.
Flower:	Tiny, white in spikes.
Fruit:	Late Summer and Fall, round or oval 1¾ inches in diameter; yellowish white or pink skin, white cottony "sweetish".
Geographic Location:	Coastal Central Florida and Southern Florida
Dormant:	Evergreen. One of a few native plants sensitive to freezing conditions.
Culture:	Sandy loam, sunny exposure, good drainage, fertilize well once a year. Little moisture, once established. Stands shearing or pruning well. Propagate from cuttings, fast growing.
Use:	The Coco Plum has become the most popular native in use in the last 15 years for general landscape plantings. Good for formal and natural hedges, informal shrub groupings. A cultivar Red Tip is in great demand.
Other:	See Medium Shrubs for "Hobe Sound Dwarf"

Coffea arabica

COFFEE

Coffea: Arabian "gawah"
    meaning wine

arabica: Latin for Arabia

Size:	To 15 feet x 10 feet, commercial plantations are pruned lower for ease of harvesting.
Form:	Upright to spreading with age. Many bearing branches are extended horizontally from the trunk.
Texture:	Medium
Leaf:	Shiny, oval, pointed, 3 to 6 inches long. Veins prominent dark green.
Flower:	Clusters of many white, ¾ inch spidery, flowers occur at leaf axils making a fine fragrant display, but short-lived.
Fruit:	½ inch red fleshy berries, 2 seeded each ½ ovoid. Berries do not mature simultaneously.
Geographic Location:	Arabia. Tropical Florida only to 32°F.
Dormant:	Evergreen
Culture:	Light rich soil. Light shade. Regular waterings. THe shallow root system calls for continuous moisture with no drought conditions, aid this by having 3 inch deep fibrous mulch over the roots. For commercial use best grown at 2000 to 3000 feet.
Use:	Rare but grown as a novelty in South Florida. A handsome plant, showy particularly when in bloom several times a year. Use as a small understory tree in hammocks and wild wooded areas. A great conversation piece.

Duranta repens

GOLDEN DEWDROP

Duranta: Named for
    C. Durantes, an Italian
    Botanist

repens: Creeping

Size:	18 feet x 12 feet
Form:	Roundish to upright form with arching branches.
Texture:	Medium
Leaf:	Opposite or whorled, oval and entire or toothed in upper half, 1 - 4 inches long.
Flower:	Pale blue, hanging loose axillary recemes about 6 inches long. Flower ½ inch in diameter with yellow eye. Some barbs.
Fruit:	Orange berries hang in masses ¼ inch in diameter and 6 inches long. Somewhat poisonous, showy.
Geographic Location:	Florida Keys, Bahamas, W. Indies, South and Central Florida. Native, subject to some winter kill, but recovers.
Dormant:	Evergreen
Culture:	Full sun, for full flowering, endures some shade. High tolerance to soil conditions. After establishing, no further care is needed. Seeds and cuttings. Rank growing.
Use:	Best use is for screens and background with smaller and ground cover plants in the foreground. Can be espaliered on fences and walls. Medium to tall informal screen. Has been used as a standard.
Other:	A white flowered variety is available.

Eleagnus pungens

SILVERTHORN

Eleagnus: Greek for olive
   and chaste tree

pungens: Sharp pointed

Size:	8 - 11 feet x 6 - 10 feet
Form:	Sprawling, weeping, improved by pruning. Long arching branch tips.
Texture:	Medium
Leaf:	3 inches long x 2 inches, oblong ovate or globose. Alternate dark green shiny above, silvery with brown dots under.
Flower:	Inconspicuous, Fall, fragrant.
Fruit:	Inconspicuous, elliptical, April, rusty brown.
Geographic Location:	Japan, China, all of Florida.
Dormant:	Evergreen
Culture:	Sun, tolerant of soils, medium drainage, moisture low. Adapts readily to pruning. Pests: spider mites. Fast growing, fertilize once a year. Survives all this state's low temperatures.
Use:	As a sheared single specimen, as an espaliered subject on concrete or stucco wall facer. At top of large scale retaining walls as screen or buffer material in shrub groups or as a hedge.
Other:	E. fruitlandi — leaves same but larger.

⌐rewia occidentalis

LAVENDER STAR FLOWER

Grewia: After Grew

occidentalis: West or Western

Size:	6-10 feet high. Sometimes equal spread, also higher.
Form:	Sprawling, loose and fast growing.
Texture:	Medium
Leaf:	Ovate-oblong, finely toothed to 3 inches long. Simple and alternate.
Flower:	1 inch wide, starlike, lavender pink with yellow centers. Blooms in late Spring with random flowers all Summer.
Fruit:	Matures orange berries in leaf axils to ¼ inch diameter in Summer and Fall.
Geographic Location:	South Africa, adapts well to warmer California.
Culture:	Pinching and pruning encourages more compactness. Requires sun and water as well as good drainage. Wind resistant. Encourage further blooming after late Spring by light pruning.
Use:	As a bank cover, if vertical growth is discouraged. Its natural flat habit of growth contributes to its popularity as an excellent espalier subject.

Hamelia patens

FIREBUSH

Hamelia: For Hamelia
Duhamel, early
French botanist

patens: Spreading

Size:	10 feet x 6 feet
Form:	Tall and graceful, but rather loose and open.
Texture:	Medium
Leaf:	3-6 inches long, whorled, elliptic-ovate, acuminate to apex. Young leaves are reddish from tomentum.
Flower:	Tubular, red, showy, in terminal forking cymes
Fruit:	Berries ¼ inch diameter, red to black, edible
Geographic Location:	Native hammocks, South Florida, coastal areas in central zone. West Indies, Tropical America, understory.
Dormant:	Evergreen. Leaves brown off in cold weather.
Culture:	Generous fertilizing will make this wildling present its best side. Sandy, well-drained soil, shady to dappled sun. Propagate by cuttings. Light pruning.
Use:	Excellent tall background subject in shrub groupings. Prime attractor for birds. As a large specimen with a background of trees it could surprise you by its grace and color. USE IT MORE!

Ilex vomitoria

YAUPON

Ilex: Hollies, an old name

vomitoria: From Indian
   lore, emetic

Size:	25 feet x 10-15 feet
Form:	Upright, irregular shrub or small tree. Compact spreading by stolens, forms colonies.
Texture:	Medium to fine
Leaf:	Alternate, gray-green, elliptic or ovate 1-1½ inches long, finely serrate/crenate margins.
Flower:	Inconspicuous, unisexual (males and females on separate plants).
Fruit:	Drupe, glistening red or yellow. ¼ inch in diameter.
Geographic Location:	Florida to Texas, Arkansas to Virginia, Native to North and Central Florida swales behind ocean dunes and hammocks.
Dormant:	Evergreen, cold-hardy.
Culture:	Sun or shade. Soil-tolerant, medium drainage, medium fertility, tolerates pruning. Salt-resistant, medium growth. No pests.
Use:	Sometimes pruned in "lollipop" standards and used in green islands in parking lots, also in median strips. Pruning can be used to shape into small multi-stemmed trees with character.
Other:	Ilex vomitoria 'Nana', a dwarf.

Lagerstroemia indica

CREPE MYRTLE

Lagerstroemia: For Myrtle
von Lagerstroem

indica: Indica of the Indies

Size:	15-25 feet x 5-15 feet spread
Form:	Upright and open or rounded. Multi-trunked with dense branching, vase-shaped.
Texture:	Medium
Leaf:	Oval 1-2 inches long, deep glossy green, Fall leaf color: yellow.
Flower:	In crinkled, crepelike, slightly conical clusters 6-12 inches long at ends of branches, smaller clusters form lower down on branches.
Fruit:	Brown woody capsules
Geographic Location:	Asia, naturalized in Florida
Dormant:	Deciduous until Spring, freeze-tolerant.
Culture:	Sun, fertilize moderately, deep watering infrequently, drought resistant. Cut back 12-18 inches to promote new flowering wood. Pests: Mildew when grown in shade.
Use:	As a tree, it is often used in line on a repetitive linear rhythm on 12 to 15 foot centers. Also in pairs at gateway entrances as a formal accent. May be pruned annually to maintain shrub size. Colors range from white to a large selection of pinks, reds and lavenders. Will present a fine tree shape if suckers are removed and the plant is limbed up as a tree.

Ligustrum japonicum

SOUTHERN WAX PRIVET

Ligustrum: Classic name
    of privet

japonicum: of Japan.

Synonym: Ligustrum lucidum

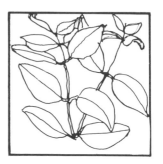

Size:	8-15 feet x 5-10 feet wide
Form:	Spreading to compact, globe-shaped
Texture:	Medium
Leaf:	Opposite broad-ovate nearly flat to 2-4 inches long, dark green, leathery. Glossy surface, lighter beneath.
Flower:	White, lilac-like odorous in terminal panicles in Spring, 8-9 inches long.
Fruit:	Berries blue-black, few in Fall. Rare.
Geographic Location:	Japan, freeze-tolerant in all of Florida. NOTE: TEXAS CALLS THIS SPECIES LIGUSTRUM TEXANUM.
Dormant:	Evergreen
Culture:	Sun or shade, good drainage. Medium fertility, medium moisture. Rapid grower, stands some pruning and shearing. Salt resistant. Was grafted for longevity until 1970. Pests: Scale, white fly and nematodes.
Use:	In the 60s and 70s Ligustrum was in great demand as a small tree: 12 feet x 12 feet, or even less. As a hedge it is still in great demand despite a current leaning toward the informal design patterns. It will always be a good background screen.
Other:	Ligustrum 'Recurvifolium';very similar in most ways with curled and recurved leaves.

Murraya paniculata

ORANGE JASMINE

Murraya: For J. Murray, a
botanist of the
18th century

paniculata: With flowers in a
loose pyramid

Size:	20 feet x 15 feet
Form:	Slender, upright, yet with a dense round head in maturity.
Texture:	Fine
Leaf:	Pear-shaped leaflets up to 1½ inches long, dark green, glossy, densely leafed.
Flower:	White, orange blossom scented, 5 petaled.
Fruit:	Red, oval to ½ inch long, attracts birds
Geographic Location:	Asia, Australia, Tropical South Florida
Dormant:	Evergreen
Culture:	Moderate fertility, moderate moisture, well drained sandy soil. Fertilize twice per growing season. Pests: Scale, nematodes, white flies and sooty mold.
Use:	Develops into large specimen globe-sheared accents. Also, a year-round good hedge from 3 feet to 6 feet. Generally in South Florida old specimen plants do grow 10 to 12 feet and, if moved, should be root-pruned in advance.
Other:	Variety Lakeview; A medium texture with larger leaves.

Myrsine guianensis

MYRSINE

Myrsine: Greek word
for Myrtle

guianensis: A plant known
to the Guianas

Size:	Often reaches 25 feet, normally 10-15 feet x 6-8 feet
Form:	Dense vertical growing head
Texture:	Medium
Leaf:	Alternate 2½-4 inches obovate-oblong to elliptic, tending to cluster near the ends of the stem, entire with recurved margins.
Flower:	Flowers are in umbels clustered along branches below leaves, are small and inconspicuous.
Fruit:	Berry is black with a thin flesh surrounding one large seed.
Geographic Location:	Native, Florida hammocks, South Florida, American Tropics, also adjusts to wet conditions with Wax Myrtle and Buttonwood.
Dormant:	Evergreen. Cold-tolerant to 28°F.
Culture:	Adjusts to variable soil conditions, moderate drainage, usually sunny or high shade, drought-resistant. Moderate growth. Pests: None. Young plants can be moved. Propogate by seeds.
Use:	Its strong vertical growth can give it use as an accent when needed. In native shrub groupings this one will always have a use. Excellent as a salt-tolerant dune plant. Myrsine has been sheared for a formal effect in adverse growing locations near the sea.

Nerium oleander

OLEANDER

Nerium: Greek name
    for Oleander

Oleander: With leaves like
    an olive

Size:	20 feet x 16 feet
Form:	Fountain-shaped, many stems flaring from base. Sometimes woody at the base.
Texture:	Medium
Leaf:	Slim, pointed, stiff, and dark green. To 8 inches long.
Flower:	Showy in clusters, to 3 inches in diameter. Petals usually white, pink or rose, sometimes double.
Fruit:	Pods are slender, 6 inches, brown
Geographic Location:	Southern Europe to Asia, most of Florida
Dormant:	Evergreen
Culture:	Any soil, any amount of fertilizers. Light watering, fast growing. Stands pruning with some loss of blooms. Fumes from burning oleanders are toxic, and the sap is poisonous. Easily propogated by cuttings. Pests: Scales, witches broom and caterpillers.
Use:	Large buffers as between residential and commercial areas. Quite tolerant of seaside salt conditions, usually an inexpensive plant used by developers because of its volume of foliage for a small outlay.
Other:	Nerium oleander 'Pink Hawaii' is an introduction and color break used at Disney World.

Ochrosia elliptica

KOPSIA

Ochrosia: From ochre, the
   yellow color of
   the flowers

elliptica: The oval form of
   the fruit

Size:	20 feet x 12 feet
Form:	Shrubby, becoming tree-like with age. Single main trunk, branches issuing evenly at shallow angle.
Texture:	Coarse
Leaf:	Large, blunt, leathery, 6 inches long arranged in pairs of 3s and 4s opposite, or whorled with many transverse veins.
Flower:	Small yellowish-white in flat clusters
Fruit:	Occurs in pairs. Lobster red, robust, drupes, 2 inches long. Poisionous to humans.
Geographic Location:	Native to New Caledonia, South Florida only. Can tolerate 28°F with no serious effects.
Dormant:	Evergreen
Culture:	Tolerant of soils. Wind-resistant, salt-tolerant, good drainage. Does best in sunny locations with fertilizer at least twice per year. Pests: Mites and scales.
Use:	Too open for good screening. Usually used as a taller plant in shrub groupings in the range of 6 feet to 8 feet when it forms single or group plants with smaller material in the foreground. Stunning foliage and fruit. Has been used as a formal accent when used or trimmed as a standard.

Pyracantha coccinea

FIRETHORN

Pyracantha: Greek for fire
   and thorn.

coccinea: Scarlet fruit

Size:	To 20 feet, usually kept to 5-6 feet x 6-8 feet
Form:	Upright and spreading multi stems with thorn-like spurs.
Texture:	Medium
Leaf:	Alternate, 1½ inch long, finely toothed, elliptical
Flower:	Small white in clusters along stems, Spring fragrant
Fruit:	Red pome, showy in Fall and Winter in clusters ¼ inch in diameter.
Geographic Location:	Japan, all of Florida
Dormant:	Evergreen
Culture:	Prune carefully to control height and spread. Sunny exposure, well drained, generous fertilizing and average watering. Vigorous grower. Pests: Red spider lace bugs and fire blight.
Use:	Fruit is a fine bird attractor. Excellent for espalier-work if maintained. Single specimens as a tree can be shaped into a king-sized bonsai.
Other:	P. coccinea 'Lalandei' most cold-hardy of many species.

Raphiolepis 'Majestic Beauty'

'MAJESTIC BEAUTY'
    INDIAN HAWTHORN

Raphiolepis: Greek for needle
    scale referring to
    flower bracts.

Majestic Beauty: Trade name
    given this chance
    seedling found by
    Charles Lee, 1969.

Size:	Can reach 15 feet, but more often seen from 4 to 7 feet.
Form:	Slightly upright to spreading habit, usually on a single stem. Branches somewhat arching, fairly dense.
Texture:	Medium to coarse
Leaf:	Alternate, 5¾ inches x 2¼-2¾ inches wide, obovate to elliptic, margin is coarsely serrated, petiole ¾ inches. Shiny, dark green.
Flower:	Large clusters of pink flowers with five to twelve petals. Each flower is ½ inch in diameter, has numerous stamens and one pistel.
Fruit:	Fruit is not conspicuous and is sterile. Blueberry-like in California.
Geographic Location:	Original seedling is from Azusa, California. Cold-tolerant from 0° to 10°F, zone seven in California.
Dormant:	Evergreen
Culture:	Full sun or light shade. Can stand fairly dry conditions, but also can stand frequent showers and waterings when first planted in lawns and flower beds. Stands pruning and some shearing. Shows tolerance to soils.
Use:	Forms a handsome background for smaller-growing Raphiolepis. Outstanding as a 5-7 foot small tree standard. As a formal accent it's hard to beat. It can more than hold its own in mixed plantings and also is a great tub subject.

Sambucus simpsonii

SOUTHERN ELDERBERRY

Sambucus: Ancient name
for Elderberry

simpsonii: Early 20th century
Florida Botanist

Size:	10-12 feet x 8-10 feet
Form:	Multistems, upright to semi-spreading. Usually open near ground.
Texture:	Medium-fine, rather lacey
Leaf:	Opposite compound with lanceolate triple leaflets graduating to double leaflets ending on odd single at top, serrated margin. Coarsely feathery.
Flower:	Creamy white, 5 parted, numerous borne in flat-topped clusters.
Fruit:	Berries are purple-black, and often collected and crushed into wine. ¼ inch diameter berry.
Geographic Location:	Native, wet to moist lowlands.
Dormant:	Evergreen, except during hard freezes, but readily comes back with flushes of new growth in Spring.
Culture:	Moist soils, fast rampant growth. Takes sun or light shade. Needs positive drainage, average watering. Recurrent sprouts from old stump or root stalks.
Use:	In large gardens they can be effective as a screen or wind break. Dense foliage can be produced by heavy Spring pruning. Fruit attracts birds. Rather brittle stems. Flowering periods in warm weather make this plant a standout in the wild garden.

Scaevola koenigii

TRUK ISLAND
    BEACH BERRY

Scaevola: Latin name after
    G. Mucius Scaevola, 6th
    century Roman hero.

koenigii: German for kingly

Size:	12-15 feet x 10 feet
Form:	Broad mound
Texture:	Coarse
Leaf:	Obovate to 6 inches long, light bright green, clustered near branch and twig ends.
Flower:	White, corolla "cut in half" in axillary panicles of one or two flowers.
Fruit:	Snowberry-like with clear white pith. Will distribute by floatation in water.
Geographic Location:	Far tropical Pacific Islands, Tropics, South Florida
Dormant:	Evergreen
Culture:	A dune fancier so has great salt tolerance. Sandy soil, good drainage, Robust grower, rank feeder. On large areas can be grown using long cuttings. Pests: unknown.
Use:	Subject known as an excellent wind-breaker. Heavy root growth and low-spreading branches (for natural marcottage) make this a successful erosion control plant on dune and river bank situations. Near ocean locations for screening parking lots and service areas.
Other:	Scaevola frutescens, similar but to 6 feet margin of leaf curled under.

Stenolobium stans

YELLOW ELDER

Stenolobium: Greek for having narrow lobes on leaves

stans: Latin referring to standing position

Size:	To 20 feet, usually 10-14 feet x 8 feet
Form:	Vase-shaped, shrubby framework. Can be a single-trunked tree occasionally.
Texture:	Medium
Leaf:	Compound pinnate with slender, light green leaflets to 5 inch pointed with serrate margin.
Flower:	Bright yellow bell-like up to 2" long in showy clusters. Summer and Fall bloomer.
Fruit:	A slim pod to 8 inches, narrow and flattened. Seed: thin-winged of very vigorous germination, springs up like weeds.
Geographic Location:	West Indies, North & South America, Escape in South Florida. Tolerant to 28°F., also Central Florida on the coasts.
Dormant:	Deciduous, but quick to leaf out.
Culture:	Almost any soil, sun, positive drainage, easily transplanted when wildings are cut back. Survives on little rain and no irrigation. Rapid growth. Control size by pruning.
Use:	More than anything else about this plant, the Fall blooming period produces the comments! Its success is when used as a background shrub in large scale group plantings.

Viburnum odoratissimum
var. Awabuki

MIRROR-LEAFED
VIBURNUM

Viburnum: Latin ancient name

odoratissimum: Latin meaning
most fragrant

Size:	To 12 feet x 7 feet
Form:	Broadly upright, dense to base. Branches broadly arching out.
Texture:	Coarse
Leaf:	7 inches long, obovate, "mirror" upper surface, dark green, lower surface light green. Coarsley dentate on upper half, not at base. New, flush growth is colored rusty orange.
Flower:	Many white, in terminal panicles, small. Fragrant in Spring.
Fruit:	Drupes red to black occasionally in Florida.
Geographic Location:	Asia, South and Central Florida, tender North
Dormant:	Evergreen
Culture:	Sun to high shade, sandy loam or muck, extra moist condition. Fast growing. In dryer situation tends to leaf-drop. 2 to 3 applications of fertilizer per year. Air layer, cuttings. Pests: None that are serious.
Use:	This is the new "too good to be true" large shrub. Be aware of the moist condition requirement. Single specimens are outstanding as accents, or as planted in regular repetitive spacing. Superb background selection for mixed shrub grouping. Probably is tolerant to shearing and pruning.
Other:	The species V. odoratissima is a tall shrub with dull leaves, fragrant flowers and is considerably less desirable.

140

MEDIUM SHRUBS

Medium shrubs included in this section are used in heights from three feet to six feet. Within these sizes are many of the most often used and colorful shrubs offered to the public.

Some of these are almost in a ground cover classification, but because of their size would be used in that manner on scales larger than the average lot size. These plants will be found used in mixed natural or informal groupings. They will be found as hedges, certain plants at the face of structures to soften the rigid lines, as backgrounds in flower borders and, beyond a doubt, might be called the workhorse of most garden projects.

A few can be found growing several feet higher than the mean of six feet, but for aesthetic reasons are usually cut or pruned back periodically to present a subject in tune with others contiguous to them.

Acrostichum danaeaefolium

LEATHER FERN

Acrostichum: Greek, refers to
　　spore arrangement
　　on leaflets

danaeaefolium: Greek, refers
　　to arrangement of leaves

Size:	Up to 10 feet with a 7 to 9 feet spread.
Form:	Multi-leaved clump.
Texture:	Coarse
Leaf:	Closely tufted and erect, ascending from a short stout root base. Leaflets are dark green with a very prominent midrib. Blades of leaflets are stiff and brittle with underturned margins.
Flower:	None.
Fruit:	Spores underside of leaflets are golden brown.
Geographic Location:	Found only in South Florida. Edge of fresh water or brackish marshes in areas that are not too wet nor too dry. Sometimes associated with buttonwoods.
Dormant:	Evergreen. Cold-proof to 26°F. Can regenerate from root-crown.
Culture:	Wetlands, but can adapt to semi-shady dry areas with irrigation and fairly heavy or peat-mixed soils. Organic fertilizer only and sparingly. No serious pests. Progagate by divisions.
Use:	Under domestic conditions this fern becomes a far less gigantic plant, growing to possibly 4 to 5 feet. Because of its scale and texture it can be used as an accent in patio plantings. Also is a handsome single specimen in a container. Its wetland background makes it perfect for use near streams and ponds.
Other:	A. aurea: almost identical but found in salt water tidal areas.

Callicarpa americana

BEAUTY BUSH

Callicarpa: reference to seed
cluster. Its beauty.

americana: native to
the New World

Size:	4 feet x 4 feet.
Form:	Round, very symmetrical.
Texture:	Medium to coarse.
Leaf:	Ovate, 5 inches long, serrate margin, soft green veins raised on underside, opposite arrangement.
Flower:	Pinkish lilac, flat-topped cyme, flowers miniature and terminal.
Fruit:	Berry-like, purple, 1/8 inch diameter, 40 to 50 clustered, Fall. Fruit occasionally found white.
Geographic Location:	Native, hammocks or pinelands throughout the State.
Dormant:	Short winter dormancy. Leafed-out in February.
Culture:	Prefers sandy hammock soils, well-drained. Dappled sun or shade. Organic fertilizer only. Propagation by seed. Can be cut back in late Fall. No pests.
Use:	This is a patrician of the wild garden for form, flower and fruit. Use it in shrub groupings as an intermediate shrub. A great attractor of birds. Designers need to build a market. Can be transplanted from the wild.
Other:	White flower and fruit occasionally found.

Capparis cynophallophora

JAMAICA CAPER

Capparis: Greek for caper,
pickled buds

cynophallophora: Latin for
penus-bearing

Size:	To 20 feet, usually 6-10 feet x 3-5 feet.
Form:	Upright to spreading. Branches turn into strong multiple secondary trunks vertically.
Texture:	Medium to fine.
Leaf:	Alternate, elliptic to obovate, 2 to 4 inches long, tip often notched. Top shiny, bottom silvery.
Flower:	Terminal, in clusters of 3 to 10 flowers, ¾ inch in diameter, white petals with stamens over 2 inches long purple with yellow anthers.
Fruit:	Slender cylindrical pods 3-8 inches long. Seeds imbedded in a scarlet pulp.
Geographic Location:	Native in hammocks South Florida. West Indies, Mexico, South and Central America.
Dormant:	Evergreen
Culture:	Cold-tolerant at 28°F, no damage. Sun to deep shade. Good drainage. Drought-resistant. Reacts beautifully to fertilizer. Propagates from seed. No serious pests.
Use:	Under nursery conditions, the subject has developed a strong upright habit with a dense leaf cover. This small tree is a wonderful addition to our palette of Florida understory material, especially in backgrounds.
Other:	C. flexuosa: a rambling semi-vine. Not attractive.

Chrysobalanus icaco

HOBE SOUND DWARF
   COCOPLUM

Chrysobalanus: Greek,
   meaning golden acorn

icaco: Arawak Indian
   for cocoplum

Size:	5-6 feet x 12 feet spread or more.
Form:	Low ground hugger, but rising to dome-shaped center, densely compact.
Texture:	Medium
Leaf:	Smooth, firm, elliptic-obovate; 1½-3 inches long, dark shiny green.
Flower:	White clusters in axillary cymes, small.
Fruit:	Pink or white obovoid, 1-1½ inches long. Flesh: sweetish, cottony, edible. Preserves and jelly.
Geographic Location:	Beach association, bayheads. South and Central Florida.
Dormant:	Evergreen
Culture:	Sandy soils, prefers sunny exposure, slow-growing. Can tolerate some poor drainage. Easily sheared and cut back. Responds to fertilizer.
Use:	Best used as a ground cover, but needs some cutting back after 3 to 4 years. Is fairly salt-resistant. Probably our most popular native shrub and, if available, is preferred for use over many exotics.
Other:	This dwarf may be a variant of the species, given dune conditions.

Codiaeum variegatum

CROTON

Codiaeum: Greek for head.
   Leaves used as
   crown garlands

variegatum: multi-colors in
   the leaves

Size:	12 feet x 10 feet, usually to 5 or 6 feet.
Form:	Rounded, mostly leggy at the base.
Texture:	Usually coarse.
Leaf:	Variable in color and shape. Thick, smooth, short-stemmed, 3 to 10 inches long.
Flower:	Small flowers in axils of leaves on 6-inch long narrow spikes.
Fruit:	Capsule, small, globose, 3-parted.
Geographic Location:	Fiji to Australia and Malaya.
Dormant:	Evergreen
Culture:	Non-selective for soils. Shade or part-sunny. Bright sun burns some broad-leafed types, tolerates shearing. Positive drainage. Moisture. Not cold-hardy. Easily propagates by cuttings or air-layers.
Use:	Enjoyed periods of popularity, but is handicapped by great tendency to develop legginess with age. Its boundless and extravagant leaf coloring is often overpowering for some uses. Is best used in an informal grouping. Not recommended for hedges.
Other:	Endless.

Eranthemum pulchellum

BLUE SAGE

Eranthemum: Greek for
  lovely flower

pulchellum: beautiful little one

Size:	5 feet x 3 feet. *Note:* Plant grows to 8 to 9 feet by end of bloom period. Should be cut back to 2½ feet at this time.
Form:	Strongly upright with irregularity and some openness.
Texture:	Medium
Leaf:	Oval, opposite, 4 inches long, deep green when shade-grown, shallow-toothed margins, sunken side veins.
Flower:	On narrow spikes 1-3 inches long above prominently-nerved, bracts tubular, blue corolla to 1 inch long.
Fruit:	Small capsules.
Geographic Location:	India
Dormant:	Evergreen
Culture:	Sandy soil of good drainage with moderate fertility. Best grown in shade to dappled sunlight. Maintain moisture. Propagate by cuttings, fast-growing. Cut back after blooming. (See note under "Size").
Use:	The blue sage presents the deepest cerulean blue flower color in subtropic Florida. Since it is a good bloomer in shade it fills a definite landscape need. Its post-blooming ragged habit requires its cutting back.

Eugenia uniflora

SURINAM CHERRY

Eugenia: for Prince Eugene of
Savoy, patron of botany

uniflora: single-flowered

Size:	25 feet x 20-25 feet. Usually 3 to 6 feet.
Form:	Typical small tree shape, either single or multi-stemmed. Because of many small twigs is a rather compact shrub.
Texture:	Medium to fine.
Leaf:	Oval, pointed, glossy, 2 inches long. Highly aromatic, dark green, new foliage is shiny red.
Flower:	Small, white and fragrant; many stamens, ½ inch in diameter.
Fruit:	8-ribbed, red-colored, cherry-sized, edible. Large stone or pit, 1¼ inch diameter. Spicy, tart.
Geographic Location:	Brazil. Below freezing temperatures as 28-32°F for 2 hours can be tolerated without serious damage.
Dormant:	Evergreen in Central and South Florida.
Culture:	Soil tolerance, positive drainage. Can stand drought conditions. Fertilize 3 times per year. Pests: caterpillars and scales. Stands frequent shearing. Propagate by seed. Fast-growing.
Use:	As a free-growing natural small tree, it is hard to beat. is commonly used as a hedge from 3 to 5 feet tall. Birds are daily visitors during summer fruiting season. Once established, no further maintenance is necessary.

Galphemia gracilis

THRYALLIS

Galphemia: old name in Greek

gracilis: for graceful form
and flowering

Size:	Maximum to 9 feet, but more often 3 to 5 feet.
Form:	Compact, very fine twiggy growth. Rounded.
Texture:	Fine to medium.
Leaf:	Opposite. Light green, ovate, foliage turns bronze in late Fall and Winter. 1 to 2 inches long.
Flower:	Showy yellow, ¾ inch in diameter, in terminal panicles.
Fruit:	Capsule, small, 3-seeded.
Geographic Location:	Tropical America. Can be damaged at 28°F. Central and South Florida and East Coast to Cape Canaveral.
Dormant:	Evergreen
Culture:	Full sun for best habit and flowering. Positive drainage, fairly dry. Growth rate is moderate. Pests: caterpillars and mites. Only bad feature is brittleness. Needs some pruning as defense against legginess.
Use:	Loose, open plants require severe shearing to form a dense head. A dominant color note in the landscape which can be used with blue plumbago in mass groupings. Can be used well as a low hedge or border plant.

Gardenia jasminoides

GARDENIA

gardenia: for Dr. Alexander
    Garden, physician,
    Charleston, S.C.

jasminoides: Jasmine-like

Size:	8 feet x 6 feet.
Form:	Rounded, either dense or loose.
Texture:	Medium
Leaf:	Pointed oval, glossy green, to 6 inches long. Opposite arrangement.
Flower:	Handsome, white, double to 5-inch spread. In Spring to early Summer. Valued especially for fragrance.
Fruit:	1½ inches. Ridged in orange. Rarely produced.
Geographic Location:	China. Hardy to 20°F.
Dormant:	Evergreen
Culture:	Sunny only. Medium to high fertility. Good potting soil. Moisture. Medium to fast growth. Pests: aphids, scales, mealy bugs, white fly, nematodes. Always buy grafted stock.
Use:	In spite of the many pests, the gardenia ranks high in popularity. Necessary oil pesticide can discolor light brick or stucco backgrounds, so keep this plant in open areas. Often best used as a tall or low standard in rather formal, symmetrical usage.
Other:	Mystery, Amy Yoshioka, Belmont, Hadley, etc.

Hibiscus rosa-sinensis

HIBISCUS

Hibiscus: ancient Greek
    and Latin name

rosa-sinensis: Chinese rose

Size:	15 feet to 8 feet.
Form:	Shrubby, often upright, robust, twiggy.
Texture:	Medium to coarse.
Leaf:	Dark green, broad-oval, apex acuminate, serrated leaf-margin, up to 5 inches long.
Flower:	Borne singly up to 6 inches in diameter. Large petal overlap, bell-shaped or flaring.
Fruit:	Ovoid, beaked capsule.
Geographic Location:	Asia, Hawaii, South Florida, subtropics.
Dormant:	Evergreen
Culture:	Full sun to dappled high shade to flower. Moderate moisture and fertility. Highly bred hybrids are usually short-lived (3 to 6 years). Pruning for hedge purposes eliminiates some flowers. Pests: snow scale, nematodes, aphids.
Use:	Of the many ways to use this subject such as specimens in natural shrub groupings, base planting, the time proven use as a hedge or screen is most popular. Because of the many forms, do not use more than one variety in a hedge.
Other:	There are dozens of varieties and colors in this family.

Ixora coccinea

RED IXORA, FLAME
OF THE WOODS

Ixora: for Malabar deity

coccinea: scarlet

Size:	10 feet to 4 feet. Usually 3 to 6 feet.
Form:	Upright, compact, well-branched.
Texture:	Medium-fine.
Leaf:	Oblong, narrow to 4 inches long.
Flower:	Scarlet, 4-petaled to 1½ inches wide, compact flat clusters to 4 inches broad. Blooms almost continuously.
Fruit:	Globose berry, red turning black, ½ inch diameter.
Geographic Location:	East Indies, Southern Asia.
Dormant:	Evergreen. Freeze damage at 32°F.
Culture:	The ixora profit from a light acidity in the soil, high fertility and positive drainage. Part shade to full sunny exposure with regular waterings. Stands shearing well. Requires trace elements. Pests: nematodes, aphids, scale.
Use:	Famed as a colorful hedge material from 3 to 5 feet. Can be shaped into formal columns. Hedge-spacing at 1½ feet apart. This was the hedge used at the Fountainbleau Hotel in Miami Beach for the extensive parterre gardens.
Other:	Some modern hybrids such as Singapore, Maui and Nora Grant are more popular than I. coccinea.

152

Ixora duffii

TRINIDAD FLAME BUSH

Ixora: a Malabar deity

duffii: unknown, botanically

Size:	6-7 feet x 5-10 feet.
Form:	Ovate profile, rather compact for a coarsely textured subject. Upright branching.
Texture:	Coarse.
Leaf:	Glossy green, broadly ovate, 6 x 3 inches. Opposite arrangement with a strongly revealed midrib vein on underside.
Flower:	Terminal cyme 10 inches wide with 50 to 60 florets, cruciform on slender tubes. Colored deep red-orange. Showy.
Fruit:	Not apparent in Florida.
Geographic Location:	Native of Malaya. Common in Trinidad and coastal South Florida.
Dormant:	Tropical evergreen. Slight drop below 32°F is damaging.
Culture:	Slightly acid soil of high fertility, good drainage. High shade to strong sun. Subtropical Florida only. Fertilize twice in the growing season using 6-6-6 with trace elements. Pests: nematodes, scales.
Use:	For conservative design, use it in mixed shrub groupings as an eye-catching color note. When used in mass or hedge form it loses impact
Other:	A well-known hybrid is I. 'Super King', even more popular than I. duffii.

Ixora 'Maui'

IXORA MAUI

Ixora: A Malabar deity

Maui: island in
Hawaiian chain

Size:	Unknown, a recent introduction.
Form:	Semi-upright to spreading.
Texture:	Medium
Leaf:	Opposite, obtuse, 3 ¼ inches long x 2 inches wide, medium, shiny green, new growth is bronzy-green.
Flower:	An umbellate flower cluster, 3 ½ to 4 inches in diameter, axillary as well as terminal, containing both closed bud (red-orange) to open florets (yellow-orange). Profuse bloomer.
Fruit:	Berries.
Geographic Location:	Discovered on Maui, Hawaii. Heritage is unknown. Considered the most cold-tolerant of the ixoras, but limit is unknown, as yet.
Dormant:	Evergreen
Culture:	Sunny exposure, well-drained, moist conditions. Prefers slightly acid soil. Fertilize 2 times per growing season. Well adapted to shearing, thus creating denseness. Suspected of sensitivity to nematodes. Treat before planting with a drench of Diazinone.
Use:	Splendid new color-break. Much is still to be learned on the best usage. The writer has tried it as a back-up for low ground covers where a third taller shrub could be used to back-up the ixora.

Jasminum multiflorum

STAR JASMINE

Jasminum: name of
     Arabic origin

multiflorum: many-flowered

Size:	Trimmed in hedge form: 6 feet, (as a vine, 15 feet).
Form:	Usually shaped as a mound form 2 x 2 feet to 5 x 5 feet.
Texture:	Fine.
Leaf:	1½-2 inches long, ovate, pointed, opposite. Downy stem. Midrib and secondary veins revealed on underside.
Flower:	White, axillary, stellate, 1 inch in diameter with fuzzy calyx teeth. Usually many flowers.
Fruit:	Inconspicuous.
Geographic Location:	India. Throughout South and Central Florida.
Dormant:	Evergreen
Culture:	Normally maintained as a shrub form by clipping. Tolerant to rich or sandy loam. Can tolerate some light shade or sun. Do not allow to dry out. Good drainage. Fast-growing. Fertilize twice per year.
Use:	When used as a ground cover, first consider the constant shaping necessary. Adaptable to hedge planting. Also for use in mixed shrub groupings. A very popular subject today. On mature plants, never-sheared runners are a problem in maintenance.
Other:	J. nitidum: a shrubby vine.

jatropha hastata

PEREGRINA

Jatropha: Greek, referring
to mediacal use

hastata: referring to foliage

Size:	To 15 feet, more likely at 8-9 feet x 8-9 feet.
Form:	Large semi-compact shrub.
Texture:	Medium
Leaf:	Evergreen, fiddle-shaped or irregularly lobed, up to 6 inches long.
Flower:	Conspicuous bright red, 1 inch across, in upright clusters. Flowers during most of the year.
Fruit:	Three-cornered pods which turn from green to yellow at maturity. Seeds are very toxic to humans.
Geographic Location:	Cuba, South Florida, tender at freezing.
Dormant:	Evergreen.
Culture:	Soil-tolerant, full sun or light shade. Water required for establishing, then discontinue. Propagate from cuttings and air-layering. Pests: mites and scales.
Use:	As an individual small tree which excels as an espaliered specimen on rubble walls and architectural wood screens and fences. Never any foliage burn on hot exposures.

Leea coccinea

KALET OR LEEA

Leea: a woman's first name

coccinea: crimson, referring
to inflorescence

Size:	6 to 8 feet x 4 to 5 feet spread
Form:	Multi-stemmed, compact, rounded profile, fern-like and lacy.
Texture:	Medium
Leaf:	Opposite, compound pinnately, 12 to 15 inches long divided into 3 to 5 pairs of compound leaflets, each ovate with an acute tip. Identify by a semi-winged clasp at all leaf joints.
Flower:	Red, but not especially showy. Flowers in cymes.
Fruit:	Not apparent in Florida. A berry.
Geographic Location:	Burma, coastal South Florida only.
Dormant:	Evergreen
Culture:	Soil a light loam or peaty soil. Grows well in bright light, but not hot sunlight. Does well in light shade as well. Best to allow for growth up as well as out. Let soil surface dry between waterings. Fertilize regularly. Defoliates at 28°F but recovers in time.
Use:	This shrub is often used in mixed plantings as intermediate or background planting. Its large lacy compound leaves of shiny green are, in themselves, a beautiful texture break wherever it is used.

Nandina domestica

HEAVENLY BAMBOO

Nandina: Japanese name

domestica: domestic usage

Size:	6 feet x 3½ feet
Form:	Upright and stiff canes, unbranching with lacy compound leaves.
Texture:	Fine to medium
Leaf:	Alternate two to three times compound, with leaflets 1½ inches long.
Flower:	Creamy white, small, in long terminal panicles.
Fruit:	Red berries ¼ inch diameter, in Winter.
Geographic Location:	Japan. South Florida to West Virginia.
Dormant:	Deciduous in North Florida. Foliage drops at 20°F.
Culture:	Medium fertility with high organic content. Good drainage. Moisture medium. Sunny for strong winter foliage color, otherwise shade. Head back old canes to staggered heights for rejuvenating. Pests: mites and scales.
Use:	Excellent vertical note against or in corners of walls and wood fences. As groups of 3 against pine tree trunks. Not recommended as a dominant plant in the landscape in South Florida because of our soil conditions and heat.
Other:	'Harbor Dwarf': low-growing 1½ to 2 feet. Orange-red in Winter.

Nerium oleander 'Petite salmon'

DWARF OLEANDER

Nerium: Greek name
for oleander

oleander: with leaves like
an olive

Size:	6 feet x 5 feet.
Form:	Vertical to rounded form, usually some foliage below. Tight branching habit.
Texture:	Fine
Leaf:	Narrow linear 4-5 inches x ¾ inch, dark olive green, pointed, grey-green underside.
Flower:	Showy, terminal, branching cymes.
Fruit:	Hanging brown pods.
Geographic Location:	Mediterranean region.
Dormant:	Evergreen. Hardy North to South Florida to 25°F.
Culture:	Suited to almost any soil, full sun conditions, tolerant to dry situations and salt conditions. Fertilize twice a year. Subject to caterpillars in the warm months. Spray with Sevin.
Use:	With 4 years since its introduction, the variety Petite pink has proved the hardiest in Florida. These plants are excellent hedge subjects to 3 feet high. When grown as a specimen, it excels its competition. The early notices called the petite a dwarf, however, some shearing (with loss of blooms) might be required.

Nicodemia diversifolia

PARLOR OAK

Nicodemia: unknown

diversifolia: with
    variable leaves

Size:	To 4 feet.
Form:	Loosely rounded shrubs. Scrambling if not maintained.
Texture:	Medium
Leaf:	Thin quilted leaf surprisingly similar to oak leaves, 2 inches x 1¼ inches. Opposite arrangement
Flower:	Inconspicuous
Fruit:	Rare
Geographic Location:	Madagascar. Central and South Florida.
Dormant:	Evergreen
Culture:	Suited to any soil conditions. Two applications of fertilizer per year. Robust grower. Can tolerate dry conditions. Good drainage. Few if any pests. Pruning increases density.
Use:	If any shrub deserves comment, this one does for its tremendous tolerance to harsh exposure and continued almost gale force winds. As a ground cover, do not allow plants to become intertwined. Space as much as 5 feet on centers and maintain.

160

Pittosporum tobira

TOBIRA SHRUB

Pittosporum: Greek for pitch
and seed's resin coating

tobira: Japanese name

Size:	8-10 feet x 6-9 feet. Used and maintained at 3 to 5 feet normally.
Form:	Compact, twiggy but stiff.
Texture:	Medium
Leaf:	Dark green, alternate, blunt, leathery, to 4 inches long. Whorled.
Flower:	Fragrant, white, heavy textured, ½ inch long, terminal umbels. Rather sparse in southern part of state.
Fruit:	Angled, globose, ½ inch in diameter, mostly sterile.
Geographic Location:	Eastern Asia. Hardy to Central Georgia, 20°F.
Dormant:	Evergreen
Culture:	Sunny exposure, good drainage, medium fertility. Medium moisture. Pruning starts with juvenile stage. Prefers a slightly acid soil. Subject to cottony cushion scale and aphids.
Use:	A patrician of the plant world. Well suited to hedge use, natural grouping in running beds and borders. Variety P. wheeleri is excellent in its low spreading habit as a ground cover with mulch; as a character-type specimen with a bonzai flare.
Other:	P. tobira variegatum: green and white foliage. Popular.

Plumbago auriculata

LEADWORT, SKY FLOWER

Plumbago: Latin for lead

auriculata: furnished with
ear-like appendage

Size:	As a shrub 4 feet x 4 feet (as a vine up to 12 feet).
Form:	Sprawling, dense foliage, billowy, twiggy.
Texture:	Fine.
Leaf:	Light to medium green, 1-3 inches long, fresh-looking, pointed, often in clusters of 3.
Flower:	1 inch wide in phlox-like clusters in white, but predominantly sky-blue.
Fruit:	Small burr-like capsule.
Geographic Location:	South Africa. South and Central Florida. Stands some frost, simply cut back injured foliage and branches.
Dormant:	Evergreen
Culture:	Good drainage, very little water after established, 2 applications of fertilizer per year. Full sun. Propagate by cuttings or seed. Pests: cottony cushion scale and mites.
Use:	Heavy rains will shred and beat flowers. To recover bright healthy look, shear off old surface leaves and flowers. Use in planter boxes allowing for fast growth. One of the subtropic's few blues. Good facer material.
Other:	P. auriculata alba: white-flowered.

Psychotria undata

WILD COFFEE

Psychotria: Greek for "vivifying" for medical properties

undata: Latin for the wavy form

Size:	Usually to 6 or 7 feet, but is known to grow to 15 feet.
Form:	Upright with loose branchwork.
Texture:	Medium
Leaf:	Shiny, elliptic-lanceolate, sharp-tipped, 2½-5 inches long. "Waffled" surface caused by recessed veins.
Flower:	White, inconspicuous.
Fruit:	Red drupe in panicles, Fall, ¼ inch in diameter.
Geographic Location:	Florida hammocks, understory. Subtropics and tropics.
Dormant:	Evergreen
Culture:	Sandy loam with organic fiber. Average woodland moisture requirements. Positive drainage. Prefers understory shade. Medium growth. Propagated by seed. No pests.
Use:	One among our finest natives for shade areas for use as a background plant or as an accent plant as in a colony in a bed of sword ferns.
Other:	P. sulzneri: Florida hammocks, leaves not shiny.

Raphiolepis umbellata

ROUND-LEAF HAWTHORN

Raphiolepis: Greek, needle-like
   scale, refers to
   flower bracts

umbellata: umbrella shape

Synonyms: R. japonica, also
   R. ovata

Size:	4 to 6 feet, occasionally to 10 feet.
Form:	Thick and bushy.
Texture:	Medium
Leaf:	Thick, leathery, roundish, dark green leaf. 1 to 3 inches long, slightly tooth-ed with thick revolute margins.
Flower:	White, ¾ inch across, fragrant, in dense pubescent panicles. Spring.
Geographic Location:	Japan. In Florida cold-resistant to Jacksonville.
Dormant:	Evergreen. In cold weather some leaf-color, including scarlet and yellow-orange.
Culture:	Best in sun, foliage thins a little and less flowering in high shade. Cut back top foliage to develop spread. Stands dry conditions but also tolerant of frequent watering. Pests: aphids in growing season. Use fungicide for leaf spot.
Use:	Tolerates salt-spray. Easily controlled size. A patrician among the medium shrubs. Blends well in natural groupings with R. Majestic Beauty.

Russelia equisetiformis

FIRECRACKER PLANT

Russelia: for A. Russel, an
    English physician

Equisetiformis: horse tail-like

Size:	4 feet x 3 feet.
Form:	Fountain-shaped, center vertical.
Texture:	Fine.
Leaf:	Rush-like stems ridged vertically with few leaves. Slim and wispy.
Flower:	All-year, bright red, slender, tubular, about 1 foot long in sprays, hanging like little firecrackers.
Fruit:	Small pendant capsules.
Geographic Location:	Mexico and Central America. South and Central Florida.
Dormant:	Evergreen
Culture:	Soil-tolerant. Full sun, well-drained soil. Pests: chewing insects, nematodes and mites. Propagated by earth layerings and cuttings.
Use:	May be used as a cascading effect in hanging planters on leading edge. Also effective on shore lines of ponds and streams. Tops of retaining walls.

Scaevola frutescens

HAWAIIAN BEACH BERRY

Scaevola: Latin name after G. Mucius Scaevola 6th Century Roman hero.

frutescens: many-fruited

Size:	3 to 6 feet x equal spread. Can attain 10 feet.
Form:	Loosely rounded to spreading.
Texture:	Coarse.
Leaf:	Fleshy bright green, convex, inverted ovate, rounded and notched at the tip, 3-5 inches long.
Flower:	White, in clusters of 5 to 9 are borne at leaf axils, fragrant, but each flower is comprised of only half of a full corolla.
Fruit:	Berries, white, pithy, 1½ inch in diameter.
Geographic Location:	Hawaii, South Florida subtropical.
Dormant:	Evergreen
Culture:	Any soil, full sun, likes moisture, fertilizer or also thrives on neglect. No pests, robust grower.
Use:	This subject will grow on marl land or on beach dunes. It thrives on ocean spray and sunshine, its original environment. Use it for erosion control, for an informal loose edge, a wind screen or ground cover on beach front.
Other:	See "Tall Shrubs" for S. koenigii.

Schefflera arboricola

DWARF SCHEFFLERA

Schefflera: Germanic name
    not traceable

arboricola: meaning tree-like

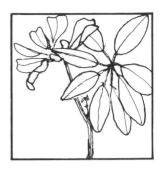

Size:	6 to 8 feet x 6 feet.
Form:	Many stems fan out into a rounded crown.
Texture:	Medium
Leaf:	Compound, palmate, 7 to 9 leaflets to 3 inches long, dark shiny green.
Flower:	Not noted.
Fruit:	Not noted.
Geographic Location:	New Zealand.
Dormant:	Evergreen. Tender tropical to 32°F.
Culture:	Extra tolerant of deep shade although it is tolerant as well of sunny conditions. Well-drained soil. Sandy soil. Cold-tolerant to 25°F. No known pests.
Use:	A recent introduction as an "interior" plant, but its out-of-doors landscape usage is amazing. Its multi-stemmed habit makes it a very fine subject for espalier work particularly for shady north exposures. As a low, informal hedge with loose appearance it is excellent.

Sophora tomentosa

NECKLACE POD

Sophora: Arabic, a flower part
   resembling a butterfly

tomentosa: Latin, covered with
   short hairs

Size:	5-6 feet. May occasionally grow to 10 feet.
Form:	Vertical to mounded.
Texture:	Medium
Leaf:	Compound, oval leaflets, ¾ inch long (one leaf compund 12-14 inches long). Leaflets are minutely hairy, silvery green.
Flower:	Light yellow corolla in terminal spikes, pea-like, most of the year, 1 foot long.
Fruit:	Brown seed pods 4½-5 inches, strongly constricted between the large seeds.
Geographic Location:	Coastal dunes and hammock margins, coastal Central and South Florida.
Dormant:	Evergreen
Culture:	Low moisture and fertilizer requirements. Sandy soil, good drainage. Salt-resistant. Provides environmental compatability. Sun preferred.
Use:	Salt-tolerant plants are few; on beach-oriented design assignments this subject can be used in beds of broad and grand scale for foliage color as well as inflorescence. Use it as a low wind break for more tender plants or ground cover.

Suriana maritima

BAY CEDAR

Suriana: named after an 18th
century French botanist,
Joseph D. Surian

maritima: Latin meaning
"of the sea"

Size:	To 20 feet, but normally 5 to 8 feet x equal spread.
Form:	Starts upright with branch tips gently arching out forming a handsome rounded to stratified specimen.
Texture:	Medium
Leaf:	Velvety spatula-shaped, to 1-1½ inches long, gray-green leaves are crowned closely at branch tips. Lower leaves turn yellow and drop.
Flower:	Inconspicuous, cup-shaped near tips, butter-yellow, 1 inch in diameter.
Fruit:	Seeds are held in a 5-pointed calyx. Good germination can be expected in the seed. 16 days to germination.
Geographic Location:	Endemic to South Florida, Caribbean, Central America, and the Bahamas. Coastal to Tampa and Cape Canaveral.
Dormant:	Evergreen
Culture:	Beach dunes, sunny, semi-dry, subject to severe storm conditions on Atlantic as well as Gulf shores in its habitat.
Use:	The state of the art re: growing and propagating is in its infancy for suriana. Nursery stock is limited to 2 nurseries where only quart and gallon containers are available. Suriana is a handsome subject limited to beach areas. It is our hope that the market demand will spark a build-up of suriana stock.

Tabernaemontana divaricata

CRAPE JASMINE

Tabernaemontana: unknown

divaricata: spreading at an
    obtuse angle

Size:	To 10 x 6 feet.
Form:	Symmetrical spreading shrub of pleasing form.
Texture:	Medium
Leaf:	Opposite, oval or lance-shaped, pointed, 6 to 8 inches long. Shiny. Milky sap.
Flower:	White, to 2 inches in diameter, ruffle-edged, clustered, very fragrant especially at night. Early Spring.
Fruit:	Orange colored pod; fruit seeds attached to a red pulp.
Geographic Location:	India, South Florida. Subtropic only. 30°F.
Dormant:	Evergreen
Culture:	Soil-tolerant, sun or partial shade, good drainage, welcomes fertilizer. Pruning to control shape is recommended. Moderately fast growth. Pests: white fly, scales, mites and nematodes.
Use:	The skeletal arrangement of branch-work makes this subject suitable for the shrub border planted at 5 feet on centers. Also a good selection for backgrounds and closures.
Other:	T. divaricata Flore-plena: double flower.

Viburnum suspensum

SANDANKWA VIBURNUM

Viburnum: Latin name for wayfaring tree

suspensum: Latin for the semi-pendant habit of blooms

Size:	4-6 feet x 4-6 feet spread.
Form:	Erect with arching branches.
Texture:	Medium
Leaf:	Opposite, oval, shining above, paler beneath, 4 inches long.
Flower:	Small tubular, pinkish white, clustered, fragrant, drooping, Winter-Spring, 1½ inches in diameter in panicles.
Fruit:	Red, subglobose drupes. Rare in Florida.
Geographic Location:	Asian islands: Liu-Kiu Island. North, Central and South Florida.
Dormant:	Evergreen. Cold-tolerant throughout Florida.
Culture:	Sun, but also shade-tolerant. Rich black sandy loam, good drainage. Grows better in Winter in South Florida than in Summer. Good irrigation. Medium growth. Pests: aphids, mites, white fly, nematodes.
Use:	Formal: low-hedge subject, 2-3 feet hedges in high shade, foundation plantings. Shrub in natural groupings and as sheared specimen as 5 x 3 foot column for accent.

GROUND COVER

Ground Cover beds are usually used as a blanket in curvilinear-formed areas which serve to soften rigid shapes such as corners of buildings, sidewalks, roadways and earth mounds. Use a hose to develop a pleasing shape. There is no maximum size for the heights of ground covers. What is important is the scale of the area. Large areas on highways can have used plants as large clumps of six foot high Pampas Grass in this manner where fifty or several hundred plants make up the bed.

At the residential level, ground covers involve smaller growing plants such as; foregrounds use the dwarf or prostrate Junipers in staggered rows backed up by medium growing plants such as Raphiolepis indica 'Alba' also planted staggered for three or four rows. Behind these, one can use medium growing shrubs such as Ixora Maui or even Ligustrum and a shade tree.

Spacing will vary according to the potential spread in two or more years from the original spread at time of planting. The open space between plants is usually covered using shredded cypress bark, two to three inches deep, pine needles, or pine bark mulch to maintain moisture and discourage many weeds.

Ground Cover areas may be uninterrupted carpets of the same plant or one may introduce one or three accent plants, a single such as a Bird of Paradise specimen, or possibly three Fountain Grass clumps in a triangle.

Finally, one should be cautioned that there is no savings in switching to Ground Covers as opposed to grassing an area. Sod is in the area of 15 to 25 cents per foot when layed, whereas Ground Covers may cost four to five times as much when measured by the square foot.

Agapanthus africanus

BLUE LILY OF THE NILE

Agapanthus - Greek for lovely flower

africanus - Latin for the origin, Africa

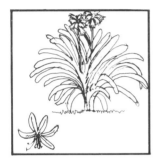

Size/Form:	20 inches x 20 inches, Roundish Clump
Texture:	Medium
Leaf:	Strap-like, "v" shaped in cross section, 20 inches long. Rising from a central rosette of tubers, dark green.
Flower:	Sky blue clusters of funnel-shaped flowers rising 6 to 12 inches above the foliage.
Fruit:	Small, light brown capsule.
Geographic Location:	Africa, South Florida 24-31°F for a short time.
Dormant:	Evergreen in Southern half of Florida.
Culture:	Likes light shade to full sun, considerable moisture and rich soil, good drainage. Planted 20 to 24 inches on center as a ground cover. Break up clumps in 6-10 years. No pests.
Use:	Often used as a marginal bed or grouping on edges of ponds. Eye-catcher because of its sky blue color. Try using it as a ground cover with a facer of Blue Daze. Used as a tubbed group in patios. Mulch heavily to protect roots.
Other:	Albus — a tall striking white cultivar.

Alocasia cuculata

CHINESE TARO

Alocasia - name made from
colocasia

cuculata - Latin for hooded;
refers to the flower.

Size:	2 feet x 18-20 inches
Form:	Upright stem with 6-8 leaves fanning out.
Texture:	Course
Leaf:	Shiny green rounded, heart-shaped from 6 inches - 12 inches in length. Stems are long. Base is a thick starchy partly exposed root.
Flower:	Rare
Fruit:	Rarely
Dormant:	Evergreen
Geographic Location:	Florida East Coast, South of Palm Beach. Hawaii, South Pacific.
Culture:	Full or dappled shade. Moist, sandy, fertile loam. Welcomes organic fertilizer and fibrous mulch. Multiples by offsets. Pests - Red spider.
Use:	Makes an excellent facer for the mixed border. Can be used as an accent (of several plants) within a ground cover bed. Also can be a quality pot plant for protected patios and terraces.

175

Asparagus densiflorus 'Sprengeri'

ASPARAGUS FERN

Asparagus - Greek for sprout

sprengeri - a man's name

Size:	2 feet x 30 inches
Form:	Spreading fountain shape
Texture:	Fine
Leaf:	Flat, needle-like, 1 inch long, light bright green, 3 or more together. Minute spines.
Flower:	Fragrant, pinkish white, small.
Fruit:	Globose berry-like, brilliant red ⅓ inch diameter, 3 seeds.
Geographic Location:	South Africa
Dormant:	Evergreen, cold tolerant to 24°F.
Culture:	Any well drained soil, full sun to dappled shade. Good soil optional. Propagated by seeds and divisions of tubers.
Use:	Popular as a facer material in mixed shrub groupings; also in ground cover beds. Adapts well as a "cascade" plant on walls and raised planters. Makes excellent material for hanging baskets.
Other:	A. Meyeri. Good looking as an individual pot plant. A bit more formal in usage than A. sprengeri. 1½ feet x 1½ feet.

Aspidistra elatior

CAST IRON PLANT

Aspidistra - Greek for stigma

elatior - taller

Size:	Leaves up to 20 inches long. Mature spread to 2 ½ feet.
Form:	Gracefully curving but with stiffness. Forms a clump. Lilylike.
Texture:	Coarse
Leaf:	20 inches long x 5 inches tapered to a point. Long petiole. New leaves generated from underground stems.
Flower:	Inconspicuous, brownish shallow bells hidden by leaves.
Fruit:	Rare
Dormant:	Evergreen
Geographic Location:	China; All of Florida
Culture:	Welcomes rather poor soil, tolerant to sun or shade. Light fertilizer. Propagated by divisions.
Use:	As an accent plant use a group of 4-6 plants in a ground cover planting of Liriope. Save this plant for the most impossible shade areas under exterior stairs, breezeways under canopies, porte-cochers.
Other:	A. lurida variegata. A white and green standout, but to keep colors use poor soil.

Borrichia arborescens

TALL SEA OX-EYE DAISY

Borrichia - Named for Olaf
        Borrich 15th Century
        Danish botanist and
        medical writer.

arborescens: - tree-like

Size:	3 feet x 2 feet
Form:	Upright to roundish
Texture:	Medium
Leaf:	1 1/8 inches - 2½ inches long, gray-green, spatulate coriacious, fleshy.
Flower:	Daisy-like disc, with yellow rays, disc often larger than the rays. Low key in color.
Fruit:	Small, sharp needle-like, 4 sided achene
Dormant:	Evergreen
Geographic Location:	Native, salt water wetland in So. FL, Bahamas, and Caribbean associated with Mangrove, also coastal sands.
Culture:	Tolerates brackish conditions, diverse soils. Full sun survives extreme dry conditions of winter and early spring. Multiplies by underground stolons, cuttings and seed. No pests.
Use:	One of a mere handful of totally salt resistant plants; ranked 5-star in my book. Performs well as a low hedge near the sea. Takes shearing well.
Other:	B. frutescens. A very weedy and proliferous type which fills swales and ditches on our barrier islands.

178

Canavalia maritima

BAY BEAN

canavalia - Latin for a strong stem

maritima - Vegetation that may be immersed in sea water

Size:	10-12 inches when mature.
Form:	Dense ground cover to 10-12 inches high. Occasional small tree climber.
Texture:	Medium-coarse
Leaf:	Pinnate compound leaflets 2½ inches - 3½ inches long, trifoliate, rounded at apex.
Flower:	Pea-type racemes in colors pink to rose-purple. Spring.
Fruit:	Robust pods, woody on wire-like vines
Dormant:	Evergreen
Geographic Location:	Frontier of dunes and coastal sands. S.E. Fla.
Culture:	Dry beach sands are typical, sunny any soil. Welcomes moisture if available.
Use:	The dense mat of foliage forms a good ground cover and resists erosion by wind and surf. Tendency to go arborial near shrubs or small tree.

Cyrtomium falcatum

HOLLYFERN

cyrtomium - Greek for arching
        & merging.

falcatum - sickle-shaped

Size:	1-2 feet tall x 2-3 feet spread
Form:	Compact forming spreading arched clumps.
Texture:	Coarse
Leaf:	Dark green leathery & glossy pinnate fronds on brown stalks 1-2 feet long. Similar to holly leaf foliage.
Flower:	Not apparent.
Fruit:	Large sori scattered underside of fertile leaves.
Dormant:	Evergreen. Hardy to 25°F.
Geographic Location:	Throughout Florida, Asia, Africa, Polynesia.
Culture:	High soil tolerance. Shade grown predominantely. Protect from frost. Water moderately during dry periods, good drainage. Subject to worms & scale. Remove old or dead fronds in spring.
Use:	Excellent as a ground cover or accent plant.
Other:	C.F. 'Rochfordianum' - Excellent foliage. More popular than the species and more robust.

GROUND COVER

Dracaena thalioides

Dracaena - Greek for female dragon

thalioides - Gray, spotted stalk

Size:	2 x 1½ feet.
Form:	One or more plants forming a single clump.
Texture:	Medium to coarse.
Leaf:	Lanceolate. 12 x 3 inches, petiole 10 inches long, dark green. Prominant parallel ribbed surface.
Flower:	10 inches long spike, globular clusters. Many small pinkish star-shaped.
Fruit:	Unknown
Geographic Location:	Ceylon, Tropical Africa, Tropic Coastal South Florida.
Dormant:	Evergreen
Culture:	Rich, moist soil. Light to very deep shade. Rather slow growing, good drainage, thrives on 2 applications of fertilizer per summer. No known pests.
Use:	One of the most shade tolerant indoor and outdoor plants in use today. Places such as outdoor stairways where minimum size is needed use D. thalioides. A cluster of 4 or 5 plants make a good porch or patio subject in pots or tubs.
Other:	D. reflexa. See under Accents.

Dryopteris erythrosora

AUTUMN FERN,
WOOD FERN

Dryopteris - Dry, winged

erythrosora - Latin for red
spore cases

Size:	1½-2 feet x 2 feet spr.
Form:	Spreading habit.
Texture:	Fine
Leaf:	One of few ferns with seasonal color value, young fronds reddish, deep green in late spring and summer.
Flower:	None
Fruit:	Reddish spore cases.
Geographic Location:	China, Japan, Florida.
Dormant:	Evergreen
Culture:	Requires shade, Western (California) Garden Book claims this fern survives all zones from 3°F up. We say in Florida it probably can survive 21°F. Forest loam, can stand drought periods. Pests - None.
Use:	Although a new introduction, its Calif. ratings indicate hot weather tolerance to 109°F. We look to a ground cover introduction with great hope.

Elettaria cardamomum

CARDAMON GINGER

Elettaria - member of the
    Zingiberacaea

cardamon - a single or lone
    heart

Size:	2½ feet x 4 feet spr. usually 2 feet x 3 feet.
Form:	Elongated horizontal rectangle, but symmetrical.
Texture:	Medium to coarse
Leaf:	18 to 30 leaves, ginger scented when crushed. Small ginger-type cane 12½ inches long containing 13 leaves each 10 inches to 12 inches x 2 inches approximately.
Flower:	Not known in Florida.
Fruit:	Not known in Florida.
Geographic Location:	Asia, South Florida, tropic.
Dormant:	Evergreen. Cold hardy to 28°F. with minor leaf damage.
Culture:	Dappled shade, moderate drainage, moist condition, two applications of fertilizer per year. Protect from winds. No known pests.
Use:	This subject can be treated as an accent using 2 or 3 plants, but also, is a handsome ground cover plant under canopy trees or large shrubs. In a mixed shrub grouping discover its benefit as a border or facer in lieu of asparagus.

Epipremnum aureum

POTHOS VINE OR TARO
   VINE

Epipremnum: attached above.

aurea: Latin for a golden
   yellow veriegation

Size:	To forty feet as a vine. Vigorous horizontal growth as a ground cover.
Form:	Normally a tree climber with multi-stems tending to occasionally over power its support clinging rootlets.
Texture:	Coarse
Leaf:	When climbing the transition of leaf from young to mature is rapid. Leaves heart-shaped, pointed, variegated to 18 inches long.
Flower:	Typical aroid spathe and spandix, rare.
Fruit:	Typical spandix vertical column with seed extending out from column when it matures.
Geographic Location:	Solomon Islands. Coastal South Florida only.
Dormant:	Evergreen
Culture:	Diverse soil tolerance. Shady or sun exposure, high precipitation or extra watering is almost necessary when establishing young vines. Needs some protection from winds. No pests.
Use:	Generally used for tropical effects as a vine on palm trees. Also presents a handsome picture as a ground cover blanket if used away from trees where it would climb.

GROUND COVER

Evolvulus glomerata

BLUE DAZE

Evolvulus - shaped like a vulva

glomerata - closely collected
      together into a head.

Size:	5 inch x 24 inch spread
Form:	Dense creepers, finely twigged, rounded shrub gradually becoming a carpet-like cover.
Texture:	Fine
Leaf:	1 inch x ½ inch broadly elliptic, full green, clustered leaves in terminal rosettes, prominent mid-rib below. Puberulent underside.
Flower:	Delft blue, axillary, 5 petalled, flat 5/8 inch diameter. Profuse. Flowers close in P.M.
Fruit:	Insignificant.
Dormant:	Evergreen can survive 26°F for short time with some damage.
Geographic Location:	Desert plant from Brazil.
Culture:	Sandy loam. Light fertilizing, full sun. Watch for fungus in rainy season. Established plants do not need moisture. Does best in elevated beds. Pests: is subject to fungus in September if not well drained.
Use:	Recent addition to the palette of ground covers for South Florida. Makes a handsome facer or edger. Cascades in raised planters and tops of walls. Time will surely add to its uses. Makes a fine hanging basket. Drought wilts leaves which recover from moisture.

Gaillardia pulchella

GAILLARDIA or
BLANKET FLOWER

Gaillardia - 18th century
       French Botanist,
       Gaillard de
       Marentonneau

pulchella - Latin for beautiful
       young girl

Size:	1½ to 2 feet high.
Form:	Semi upright, loosely branched.
Texture:	Medium
Leaf:	Oblong spatulate to 4 inches long. Softly pubescent, entire to cut, pinnatly.
Flower:	6 to 15 rays each ½ to ¾ inch long; tips are yellow balance is rose grading rose purple at base. The floral rays can be found from a solid pale yellow to tricolor.
Geographic Location:	Native to Florida, Kansas, Arizona, Louisiana.
Dormant:	Year around in Central and South Florida; Evergreen
Culture:	Should re-seed itself. Is an annual. Plant in light open, well drained soil in full sun. Stands drought well, water sparingly. After a year extend the life of your plant by cutting back. DO NOT CUT BELOW THE GREEN WOOD.
Use:	Most natives are not known for splashes of color; but gaillardia is! Use it in large colonies and try to use a mix of colors. Behind the primary dunes it becomes a handsome dwarf. Liner-plants from pots will save money and quickly mature.

Gamolepis Chrysanthemoides

AFRICAN BUSH DAISY

Gamolepis - united scales or
    petals.

chrysanthemoides - like a
    chrysanthemum.

Size:	2½ feet x 3½ feet (Grows double ht. and spread shown, but needs cutting back.)
Form:	Broadly rounded.
Texture:	Medium
Leaf:	2½ inches long x ¾ inch wide coarsely feathered light bright green alternate.
Flower:	1 7/8 inch in diameter on 4 inch stem, 12-14 petals, butter yellow, profuse disc yellow.
Fruit:	Many-seeded in flat disc.
Geographic Location:	South Africa
Dormant:	Evergreen. Tolerates light frost to 28°F.
Culture:	South African plant growing in non-selective soils, full sun, occasional clipping or heading back to keep dense, drought tolerant well drained situation. Average fertilizing. Reseeds itself. Pests: Nematodes.
Use:	Colorful ground cover or for use in mixed border. In Florida. Gamolepis is a great bloomer from spring through summer and into early winter.

Gazania longiscapa

GAZANIA DAISY

Size:	18-20 inches x 18 inches
Form:	Clumping
Texture:	Medium
Leaf:	White-wooly below, lanceolate or pinnately cut.
Flower:	Daisy-like multi colored or single colored. White thru yellow to orange; lavender & pink to deep red in concentric rings. 2 inches across; closes at night.
Fruit:	Profuse seed
Geographic Location:	Africa
Dormant:	Evergreen. Tolerant to 26°F.
Culture:	Almost any soil. Good waterings; sunny locations; to extend the life of this annual, cut below seed heads, but not into brown wood.
Use:	Salt tolerant; in flower border, as a ground cover. One of the best color notes available. Seeds & divisions.
Other:	Trailing Gazania, rigens, spreads rapidly by long trailing stems.

Helianthus debilis

BEACH SUNFLOWER

Helianthus: pertaining to
   sunflowers

debilis: weak, referring to
   trailing habit

Size:	3 feet x 5 feet or more spread in irregular circles.
Form:	Low growing and robust spreader
Texture:	Medium
Leaf:	2-4 inches long alternate deltoid, irregularly dentate rough surface. Medium green.
Flower:	Pale yellow 10-20 rays, disc ½-1 inch in diameter spring & summer. Disc is dark maroon in color.
Fruit:	Seed is viable and reseeds easily.
Geographic Location:	Dunes area. Also waste lands, inland; all Florida.
Dormant:	Evergreen
Culture:	Any soil, moisture from occasional rains, well drained sunny exposure. When collecting, use seedlings or small plants on 2½ feet centers. Reseeds itself readily, biennial.
Use:	Popular as a ground cover in near-beach locations where it can reseed itself. Its relocation can surprise one, since it will leave its designated area and show up 50 feet away, so resist the temptation to use big beds. Super salt resistant.

Hemerocallis 'Aztec gold'

DWARF EVERGREEN DAY
   LILY

Hemerocallis: Beautiful
   roughened protrubence

Size:	1 foot x 1 foot
Form:	Lily-like clumps
Texture:	Medium
Leaf:	Strap-like, grows in clusters from roots that are tubers.
Flower:	Trumpet-shaped, vivid orange, 3-4 inches long
Fruit:	Rare
Geographic Location:	Beachfront, forest perimeter or sunny open lawn area.
Dormant:	Evergreen
Culture:	Prefers rich sandy loam. Periodic fertilizers at 4 to 6 times per year. Light shade to sun, well drained soil condition. Few pests. Propagate by divisions.
Use:	Superb, March to June blooming ground cover plant on 18 to 24 inch centers using heavy gallon containers. This advice is necessary: You must adhere to the need for fertilizing as per culture directive.
Other:	Most evergreen varieties that are happy in Florida are tall varieties in other colors.

Ilex vomitoria 'Nana'

DWARF YAUPON

Ilex: Ancient Latin name for
    Quercus ilex

vomitoria: Emetic

nana: dwarf

Size:	2-4 foot x 3-5 foot spread
Form:	Dense. Flattened globe, symmetrical
Texture:	Fine
Leaf:	To 1 inch long, alternate, elliptic-serrated, dark green
Flower:	Few, inconspicuous.
Fruit:	None
Geographic Location:	
Dormant:	Evergreen
Culture:	Sun or shade, soil widely varying, fairly salt-tolerant, well drained. Pests: scales & spittle bugs.
Use:	Prime ground cover material by virtue of its crisp rounded form; offers a more formal presentation than most others. Also, makes a low rigid hedge or edger.
Other:	I.v. 'Shillings', reddish tips.

Ipomoea pes-capri

RAILROAD VINE OR GOATS-
  FOOT VINE

Ipomoea: worm-like

pes-capri: foot of the goat

Size:	8 to 12 inches as a mat.
Form:	Trailing vines extending to 75 feet arising from a thick, starchy root and branching from the joints forming a thick mat on the beach.
Texture:	Coarse
Leaf:	4 inch Long, thick rather succulent obovate notched or 2 lobed at apex.
Flower:	Lavender funnel-shaped flower with pruple-red throat.
Fruit:	Leathery pod ½ inch long. Seeds ½-⅓ inch long.
Geographic Location:	Native. Sandy beaches and coastal areas
Dormant:	Evergreen
Culture:	Porous beach sand, with or without moisture. Full sun only. Do Not Overwater. Closes blooms in late p.m.
Use:	With some care this plant explodes and becomes a richer ground cover with nodes at shorter intervals. An excellent sand binder. High salt tolerance.

Jasminium nitidum

GOLD COAST JASMINE

Jasminum: Ancient name of
  arabic origin

nitidum: Shiny-leaved.

Size:	2½ to 3 feet
Form:	Clambering or climbing, semi-mounded
Texture:	Medium
Leaf:	Shining, pointed. 2-3 inch long, oblong, thick. Tips of new shoots are tinged rosy purple.
Flower:	White, rosy, when opening. Fragrant, over 1 inch long. Blooms open at nite. Continuous bloomer.
Fruit:	None
Geographic Location:	West Africa
Dormant:	Evergreen
Culture:	Good drainage, sunny location, moist rich soil. Average fertility. No enemies.
Use:	A strong grower not too easily maintained as a tight shrub. Shear often to control growth. This subject is your best bet as a hedge having a loose informal effect. Tends to become viney.
Other:	See other Jasmines

Jasminium volubile

WAX JASMINE

Jasminum: Ancient name of
　Arabic origin

volubile: climbs spirally

Size:	Rarely grown over 24 inches x 3 feet
Form:	Bushy and dense, rounded.
Texture:	Fine to medium
Leaf:	Shiny deep green, elliptic. 2 inches long. Opposite, graceful arching.
Flower:	White, 1 inch long, petals narrow & sharp. Not profuse.
Fruit:	None
Geographic Location:	Africa, South Florida
Dormant:	Evergreen. Cold sensitive at freezing.
Culture:	Rich sandy loam, well drained moisture. Adapts well to shearing to 18-24 inches as a hedge. No known pests.
Use:	Attractive ground cover. Good spacing at 4 feet on centers with mulch. Fast growing. Low hedge and divider material formally clipped or informal unsheared.
Other:	See other Jasmines.

Juniperus squamata expansa 'Parsoni'

PARSONS' JUNIPER

Juniperus: Juniper-like

chinensis: From China

Parsonii: After a Mr. Parson

Size:	1½ feet x 8 feet spread
Form:	Dense short twigs on flat, rather leafy branches.
Texture:	Fine
Leaf:	Starts with scale leaves and changes to needle-like, dark green, dense.
Flower:	Inconspicuous
Fruit:	Rare
Geographic Location:	J. chinensis is Chinese, Parsoni is a mutation.
Dormant:	Evergreen — tolerates 0°F
Culture:	Succeeds in varying soils, but not in continuously wet conditions. As a young nursery grown plant, Parsoni does not form symmetry for a year or more. Well drained soil condition. Space on 3 foot centers and mulch.
Use:	As a ground cover hard to beat. Young plants do not grow out evenly, but they fill in given two growing seasons. More upright than the two following varieties, which are excellent subjects as ground covers.
Other:	Juniperus chinesis procumbens, 3 feet x 12 feet sp. Juniperus chinensis procumbens nana, 12 inches x 4-5 feet.

GROUND COVER

Juniperus conferta 'Compacta'

DWARF SHORE JUNIPER

Juniperus conferta: Densley compact.

Compacta: Crowded, pressed
together. Dwarf.

Size:	6-8 inches x 4 feet spread
Form:	Prostrate carpet-grower
Texture:	Fine
Leaf:	Gray-green with white lines on top
Flower:	Inconspicuous
Fruit:	Rare
Geographic Location:	J. conferta is Japanesse. Var. compacta is a mutation.
Dormant:	Evergreen. Tolerates 0°F
Culture:	Prefer sandy loam. Soil — good drainage; low fertility. Moisture — low. Do not prune. Rapid growth. Can stand some shade. Pest Problems: red spiders.
Use:	Salt resistant. Effective ground cover on slopes of berms and banks. Useful as foregrounds for taller plants. Will cascade in planters. For beach use plant on leaward side of dunes.
Other:	Juniperus conferta. 12 inch x 6-8 feet, coarser than above, subject to some die-back in wet weather, probably a fungus. Juniperus conferta, Blue Pacific: more conpact than Juniperis conferta, better blue color.

Lantana montevidensis (sellowiana)

WEEPING LANTANA

Lantana: Old Italian word for Viburnum which Lan tana resembles in leaf.

montevidensis: For Argentine city

Size:	Variable x 4-6 feet long (trailing)
Form:	Trailing vine-like shrub
Texture:	Medium
Leaf:	Ovate 1 inch long and rough, opposite, twigs square, dark green foliage, aromatic.
Flower:	Lilac 1-1½ inch wide in clusters, almost continuous bloomer.
Fruit:	Small drupe, rare
Geograpic Location:	South America — South Florida
Dormant:	Evergreen
Culture:	Ground cover, fast growing, well-drained sandy loam. Sun or shade, but sun to bloom. Tolerates freezing temperatures to 20°-30°F. Pests: Nematodes.
Use:	An old favorite for ground cover beds, blooming Spring thru Fall. In planter boxes it cascades in a grand colorful curtain. Very salt resisitant.
Other:	Lantana camara — coarse upright to 6 feet. Parent of a number of colored selections and hybrids.

Lantana ovatifolia
var. reclinata

GOLD LANTANA

Lantana: Old Italian name
for Viburnum, which
is similar

ovatifolia: Ovate leaf

reclinata: Reclined or
semi-prostrate

Size:	8 to 10 inches
Form:	Spreading, forming a carpet
Texture:	Medium
Leaf:	Light, bright green minutely waffled surface, underside pubescent. 1¾ to 2¼ inches long x 1 inch wide, keeled. Margine crenate.
Flower:	In head ¾ inch in diameter, on axillary peduncles 1½ to 2 inches long. Stems 4-angled. Flowers, 10-15 yellow about ¼ inch in diameter in flat heads.
Fruit:	Drupes subglobose
Geographic Location:	Native to South Florida pinelands
Dormant:	Evergreen
Culture:	Of easy culture and adapatble to soils. Dappled shade or bright sun. No fungus or wet season problems once established. No maintenance.
Use:	This native has been outstanding for the few years it has been made available. On a scale of one to ten, this Lantana deserves a ten for ground cover. One of the best long-blooming yellow flowers of any now on the market.

Liriope muscari

BIG BLUE LILY TURF

Liriope: For the nymph, Lirope

muscari: Refers to musky odor

Size:	12-18 inches x 12-14 inches spread
Form:	Grass-like with leaves recurving to ground.
Texture:	Fine
Leaf:	18 inches long x ¾ inches wide, dark green.
Flower:	Lilac purple on a spike.
Fruit:	Black shiny
Geographic Location:	Japan and China
Dormant:	Evergreen, Tolerates 0°F
Culture:	Slow growing. Shade loving, but stands sun. Likes ample moisture. Well drained, medium rich soil and generous use of fertilizer. Subject to slugs and snails.
Use:	Best used as a casual cover in smaller scaled areas as a foreground material backed by larger material. As a cover below a sculptured plant such as strelitzia regina, or in conjunction with a border or edger in Southern gardens.
Other:	Liriope 'Evergreen Giant' — spreads to 24 inches, tolerates full sun, preferred today over Liriope muscari. Liriope 'Websters' Wideleaf, Coarse textured, 12 inches spread, dense, does best in dappled shade. L. Evergreen Giant variegated; clear striped white and green and sun tolerant.

Malpighia coccigera

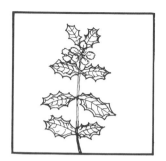

DWARF HOLLY

Malpighia: Named for M.
    Malpighi, Italian
    Naturalist

coccigera: Berry-bearing

Size:	2½ feet x 2½ feet
Form:	Dense twiggy rounded, dainty
Texture:	Fine
Leaf:	Opposite, miniature holly-like, ¾ inch long medium green
Flower:	Flaring pink, wavy petaled ½ inch diameter auxillary cymes
Fruit:	Globuse, red to red-orange ⅓ inch diameter.
Geographic Location:	West Indies — South Florida coastal areas
Dormant:	Evergreen
Culture:	Tender Tropical. Sunny to dappled shade, well drained soil, medium fertilizer, moisture.
Use:	Has been one of South Florida's best known dwarf shrubs, which has been used as a clipped hedge for edges and borders. Needs trimming or shearing to become uniform, since there is variation when produced from seed.
Other:	A number of selected varieties have been improvements of the species.

Nephrolepis exaltata

SWORD FERN

Nephrolepis: Likened to fish
    scales

exaltata: Rasied high or lofty

Size:	To 3 feet high
Form:	A symmetrical arrangement of upright, erect fern fronds, ladder-pinnate.
Texture:	Fine
Leaf:	Lettuce green, pinnae close together, tapering gradually to a dagger point.
Flower:	None
Fruit:	Spores
Geographic Location:	Native, E. Asia, Africa, Florida to Brazil.
Dormant:	Evergreen
Culture:	These natives are collected often on shady ditch banks, edges of wooded areas, hammocks, and swamp margins. Sun or shade, sandy soil often on Cabbage Palm trunks, multiplies readily.
Use:	Use sun-grown ferns for sunny locations. As a ground cover use it in large colonies; it does not blend well with Junipers and most evergreen broad-leafed ground covers. Its light bright green color is good in deep wooded areas as a highlighter.
Other:	N. biserrata, arching fronds to 3½ feet to 4 feet. N. Cordifolia, fronds erect to 30 inches long, roots creeping tubiferous.

Ophiopogon japonicus

LILY TURF

Ophiopogon: Greek for Snake's Beard

japonicus: From Japan

Size:	6 inch - 10 inch spread indeterminate.
Form:	Grass-like cover of stemless clumps. Carpet ground cover, dark green.
Texture:	Fine
Leaf:	Grassy, 9 inches - 12 inches x 1/8 inch wide curving to ground.
Flower:	Not important; hidden by foliage
Fruit:	Berry, light blue
Geographic Location:	Japan, tolerant to 0° F. all of Florida
Dormant:	Evergreen
Culture:	Good drainage, sandy soil, shade preferred. Medium fertility with humus added. Set plants four inches to 6 inches apart, rapid growth. Pests: scale. Cold tolerant to South Pennsylvania.
Use:	Especially useful under trees where grasses die out. As a ground cover on slopes it offers good erosion control when established. Often is used as border or edger facing down background of semi-dwarf shrubs.
Other:	O. jaburon vittata leaf 12 inches - 18 inches x ¼ inch. ½ inch variegated. Best used in dappled sun or shade.

Pentas lanceolata

EGYPTIAN STAR FLOWER

Pentas: Greek 5; referring to
  five flower parts

Lanceolata: lance-shaped leaf

Size:	3 feet to 4 feet (subshrub)
Form:	Sprawling leafy mound dotted with colorful blossoms.
Texture:	Medium
Leaf:	Opposite, ovate to oblong-lanceolate, pointed rough surface.
Flower:	White, lavender, pink and red, multiple tubular stars 1 inch, terminal umbels.
Fruit:	Capsule, bears many small seeds
Geographic Location:	Tropical Africa
Culture:	Evergreen: South Florida subject to damage at 31°F. Fertile black sandy soil well drained, moderate moisture. Remove old flower heads to maintain blooming period. Replant every 2 to 3 years in nematode free soil. Fertilize lightly once each growing month.
Use:	More than any other perennial Pentas is best used in drifts of several separate colors sandwiched with banks of white between. Pentas enjoys cyclic popularity — almost impossible to find or buy at times. Needs to have a renewed market at the time of this writing.

Peperomia obtusifolia

PEPEROMIA

Peperomia: Pepper-like

obtusifolia: Wide angled leaf

Size:	To 8 inches x 6 to 8 inches
Form:	3 to 4 stems
Texture:	Coarse
Leaf:	4 inches x 2½ inches dark green round or oval
Flower:	Spike 2 inches to 6 inches long, like a green rat-tail.
Fruit:	Not Important
Geographic Location:	Found in shady, moist tropical American jungles. Escape in South Florida.
Culture:	Shade to light sun. Moisture rich composted to sandy soil, high humidity preferred, good drainage. Propagates easily from cuttings.
Use:	Shade oriented ground cover usually for confined spaces. Its handicap is its brittleness: dogs and humans walking thru a bed will leave their footprints of destruction. Fine for planters as a filler, also as a pot plant.
Other:	Several weeping species are excellent for hanging baskets. There are over 70 varieties and species listed in Exotica.

Pilea serpyllacea 'Rotundifolia'

ARTILLERY FERN

Pilea: Latin for felt cap for the fruit cover

serpyllacea: Latin for thyme-leaved

rotundifolia: Tends to have rounded leaves

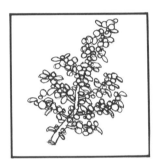

Size:	8-12 inches
Form:	Miniature tree-like, semi upright, tight, subshrub
Texture:	Fine
Leaf:	Orbicular rounded at the base, heavy lateral branching opposite leaves.
Flower:	Inconspicuous
Fruit:	Inconspicuous
Geographic Location:	Tropical America
Dormant:	Evergreen. Sub-tropic only.
Culture:	Treat as an annual or biennial. Well drained sandy loam. Light fertilizer (Osmacote is good) moisture. Pests: subject to worms. Sunny
Use:	This subject is a color break from most plants, being a bright chartreuse green. Usage as a short-lived ground cover among other ground covers as a filler is recommended. Otherwise it does a good job as an edger or even a facer material.

Pittosporum tobira 'Wheeleri'

WHEELER'S TOBIRA SHRUB

Pittosporum: Greek for pitch and seed

tobira: Native Japanese name

Size:	1½ feet x 3 feet
Form:	Flattened globe or wheel
Texture:	Medium to fine
Leaf:	Obovate to 4 inches long, thick and leathery, dark green, margin revolute, densely compacted foliage.
Flower:	Not apparent
Fruit:	None
Culture:	Evergreen, tolerant to 10°F. A selected mutant it prefers a prepared topsoil with adequate moisture with fertilizar 2 to 3 times per year. Positive drainage, add trace elements, prune or shear to unify shape as a ground cover. Subject to scale, red spider, aphids and fungus.
Use:	This will present an elegant picture as a ground cover either in a rectilinear design in a plaza, or in an informal drift pattern. Allow ample room for future growth filling with mulch or a small colorful low plant such as purslane or pilea.
Other:	P.tobira, the parent is considered a large shrub to 10 feet and is treated under large shrubs. Wheeleri is a mutant.

Polypodium phymatodes

LAUA'E, EAST INDIAN WART FERN

Polypodium: Latin having
   many feet

phymatodes: "Warty" referring
   to indented spore cavities
   underside of leaf.

Size:	To 3 feet on individual leaves.
Form:	Brown scaly root-stock, creeping stolonate with leaves occuring irregularly.
Texture:	Medium to Coarse
Leaf:	Dark green flat fronds, oblong lobed as in Philodendron selloum. 2 feet long to 1 foot wide. Prominant brown midrib.
Flower:	
Fruit:	Spore cavities on underside of leaves show on upper surfaces as warts single rows either side of midrib on lateral and terminal lobe.
Geographic Location:	Hawaii, Polnesia and Old World Tropics
Dormant:	Evergreen, semi-resistant to light freezes with some temporary damage. Recovers well.
Culture:	Light shade, well drained fibrous soil or mulch or even black muck. Stolons branching and marching in all directions. Sometimes found climbing tree-trunks.
Use:	Probably the most luxurious shady ground cover in the book. Can be used in close proximity to the leather fern with success. A good hanging basket subject. A good facer material that resists foot wear.

Portulaca oleracea

PURSLANE, SATIVA

Portulaca: Like the Portulaca

oleracea: Edible, esculent
    edible vegetable

Size:	4 inches x 10-12 inches
Form:	Prostrate
Texture:	Fine
Leaf:	Obovate ½ to 1½ inch diameter.
Flower:	Petals 5, 1 inch diameter. Colors range from salmon red to orange to yellow.
Fruit:	Capsule, small.
Geographic Location:	Tropical America. South Florida subject to frost.
Dormant:	Evergreen
Culture:	No special soil, well-drained, sunny, light fertility, fast growing cuttings root readily. Pests: none serious. Rather short lived. Should have regular maintenance.
Use:	Colorful ground cover best used as a bedding plant in foregrounds. Adapts to hanging basket type use. Color clashes must be avoided when using bicolor or tricolor combinations.

208

Pyracantha 'Lodense'

DWARF FIRETHORN

Pyracantha: Greek for fire
  and thorn

lodense: A contraction of low
  and dense

Size:	To 2½ feet a generous spreader
Form:	Grows into tight mounds as it spreads
Texture:	Fine
Leaf:	Closely set leaves in miniature, gray-green, twiggy compact form.
Flower:	Occasional heads in small scale of creamy white.
Fruit:	Sparse crop of small orange berries often hidden by foliage.
Geographic Location:	Florida
Dormant:	Evergreen cold tolerant to 20°F.
Culture:	Adapts to almost any soil. Sunny, well drained, rank-grower occasional fertilizer. No pests.
Use:	A good, fine textured ground cover, best spaced on four feet centers in heavy mulch. Also, is an attractive facer on border type plant in natural informal plantings. A small hedge with maintenance. Ideal planter box subject which will cascade.

Raphiolepis indica

INDIAN HAWTHORN

Raphiolepis: Greek for needle scale, refers to flower bracts.

indica: From India

Size:	3-4 feet x 4 feet
Form:	Spreading, open, irregularly branched
Texture:	Medium
Leaf:	Alternate, thick, leathery pointed, margins are entire.
Flower:	Terminal panicles, pink or white ½ inch in diameter, 5 petals.
Fruit:	Black drupe 3/8 inch diameter.
Geographic Location:	China, India for all of Florida
Dormant:	Evergreen
Culture:	Sun or shade. Tolerant to soils, good drainage average moisture. Slow growing. One of our most salt-resistant shrubs. Somewhat subject to scales.
Use:	Suitable for generous use on primary dunes. Tremendous show of color when used as a dominant ground cover for spring. Good border or facer plant and needs little maintenance. Slow-growth equals easy maintenance.
Other:	Raphiolepis umbellata. 4-6 feet, white flowers ovate foliage.

Rhododendron obtusum

KURUME AZALEA

Rhododendron - Greek;
      Rhodon is rose,
      Dendron is tree

obtusum - Latin for blunt
      or obtuse.

Size:	2 feet x 2½ feet
Form:	Dense spreading, wider than high.
Texture:	Fine
Leaf:	Dull green ¾ inch x ½ inch growing compactly tapering to blunt point.
Flower:	Varies in color according to variety, as Salmon, Pink, White and Red. Blooms sometimes hide foliage.
Fruit:	Small terminal seed (fine) capsule.
Dormant:	Evergreen
Geographic Location:	Japan. Central to North Florida.
Culture:	Shady to dappled shade, soil-good drainage; medium fertility with a fairly thick blanket of mulch. Well watered, must maintain a good acid reaction always. Pests: Lacebugs, scale, spider mites and root rot.
Use:	In South Florida in combination with Wax Myrtle as overstory and limbed-up. In ground cover beds in shade. Also as a facer material with taller azaleas.
Other:	Best for South Florida: Redwing-deep red, Duc de Rohan-medium salmon.

Rhoeo discolor 'Bermudiana'

DWARF OYSTER PLANT

Rhoeo: Tradescantia-like, old
    name.

discolor: Bicolored

Size:	12 inches x 16 inches
Form:	Clustered plants form the clump
Texture:	Medium
Leaf:	Sword-shaped, tender (brittle), succulent, 8 inch long green above, purple beneath.
Fruit:	Miniscule, not important
Geographic Location:	Bermuda, American tropics. Sub tropic South Florida
Dormant:	Evergreen — not cold tolerant
Culture:	Best grown in sandy, well-drained soils, do not overwater. Sunny, organic fertilize twice a summer. Rapid development into clumps. Pests: fungus, root rot, leaf spot. For root rot use Benlate, 1 pound per 100 gals.
Use:	When used as an edger plant at least 2 rows staggered. As a ground cover in a curvilinear bed allow for at least 4 rows deep.

Rumohra (polystichum) adiantiformis

LEATHER LEAF FERN

Rumhora: Unknown

adiantiforme: As in the form of the Maidenhair Fern

Size:	2 feet tall in a well knit mat.
Form:	Spreading and overlapping stolens and leaves
Texture:	Fine
Leaf:	Fresh light green triangular fronds. Thick, leathery, 1-3 pinnate with oblong segments coarsely toothed.
Flower:	None
Fruit:	None
Geographic Location:	South Pacific Islands, Tropics of South America, New Zealand and South Africa, Southern and Central Florida.
Dormant:	Evergreen
Culture:	Terrestrial, but prefers a fibery leaf-mold with some sphagnum moss. Grows beautifully in shade, but can tolerate some sun. Hardy to 24°F. Needs a moderate amount of water.
Use:	Forms a tightly knit bed of rusty-redwood colored stolens topped with an 18-24 inch high blend of pointed fronds in all directions, well organized. Is unusually fine ground cover in dappled shade.

Spathiphyllum clevelandi

PEACE LILY

Spathiphyllum: Greek for
    leaf-spathe

Size:	12-18 inches x 18 inches
Form:	Clump form building from vertical leaves centrally located to arching leaves in perimeter.
Texture:	Medium
Leaf:	Lanceolate, sharply pointed 4-6 inches petiole 6-8 inches long.
Flower:	White with ovate pointed, papery spallae turning apple green with age.
Fruit:	The seeds are extruded from cob-like spadix or spike.
Geographic Location:	Tropical America
Dormant:	Evergreen
Culture:	Rich sandy loam in moist humid atmosphere. Well drained, light to deep shade. Fast growth. Pests: Mealy bugs, scales and mites.
Use:	S. clevelandi cannot stand direct sunlight. An excellent tropical in beds combined with some gingers, polypodium ferns. Anthuriums peperomias and the like.
Other:	S. Mona loa, bigger copy of clevelandi, spectacular — Hybrid.

Spathiphyllum 'Wallisii'

DWARF PEACE LILY

Spathiphyllum: Greek spathe-
    like leaf
*wallisii:*Developer's name

Size:	20-24 inch spread x 12-15 inch height
Form:	Symmetrical in plan and side view, 3 to 4 layers of overlapping foliage. Slow growing.
Texture:	Medium
Leaf:	10 inches long, shiny dark green. Leaf is conversely arched between veins picturing a rippled effect.
Flower:	Small spandix slightly above crown of the plant, chalk white. Usually not profuse. Spring blooming.
Fruit:	Inconspicuous
Geographic Location:	Columbia, S.A., in Florida, sub-tropic, S.E. Coastal area
Dormant:	Evergreen
Culture:	Rich soil, light to medium shade, moist conditions. Fleshy roots, fertilize with 20-20-20 drench and foliar spray. Protect from winds and freezing conditions.
Use:	For use in shaded protected patios as a ground cover. A carpet effect is obtained by a 24 inch spacing. Works well as a facer for taller spaths. Use in shore or bank plantings, in indoor falls or pools. Usually purchased 8 to 10 per pot.
Other:	S. londonii, a larger wallisii hybrid from Mauna loa

Trachelopermum jasminoides 'Minima'

DWARF CONFEDERATE JASMINE

Trachelopermum: Greek
   combined for neck and seed

jasminoides: Jasmine like

minima: Latin minimum
   refers to size

Size:	As a ground cover not over 8 inches forming a carpet.
Form:	A dense mat, tends to get tangled if allowed to grow over 6-8 inches.
Texture:	Fine
Leaf:	2 x 1 inches elliptic with pointed tip, opposite, simple shiny deep green.
Flower:	White, 5 petaled pin wheel counter clock-wise small fragrant in leaf axils.
Fruit:	Rare
Geographic Location:	China, fairly hardy even in North Florida
Dormancy:	Evergreen
Culture:	6 hours minumum sunshine. Occasional pruning of wild tendrils. Average good soil, well drained. May need weeding the first year or more. Likes moisture. Pests: scales and sooty mold.
Other:	T. asiaticum "minimound", "red top", and Texas Longleaf.

Tulbaghia violacea

SOCIETY GARLIC

Tulbaghia: Refers to a man's
name

violacea: Latin for flower color

Size:	Up to 2 feet high
Form:	Grass-like clumps which multiply
Texture:	Medium to fine
Leaf:	Erect linear numbering 8-20 leaves, 12 inches long and 1/8 inch wide, bluish green.
Flower:	Rosy-lavender flowers, 8-20 star-shaped in a cluster on stems 1 to 2 feet long. Long season of blossom. Early spring into summer.
Fruit:	Capsule
Geographic Location:	South Africa. South Florida. Frost damage at 25°F, but quick recovery.
Dormancy:	Evergreen:
Culture:	Light sandy soil, full sun, average waterings, good drainage. Fertilizer twice in growing season.
Use:	The light lavender flower color although not smashing, makes a strong impact when planted as a ground cover in depth and quantity. Also, is an effective edger or border in a shrub grouping.

GROUND COVER

Vinca rosea or Catharanthus roseus

MADAGASCAR PERRIWINKLE

Size:	1½ to 2 feet x 2 feet spread.
Form:	Bushy, but not very dense.
Texture:	Medium
Leaf:	The glossy leaves are short-stemmed and paired with a pointed tip, 1-3 inches long.
Flower:	Showy, five-parted, rose-purple or white with or without a red throat. 1 to 1½ inches in diameter.
Fruit:	Not common, inconspicuous pod.
Geographic Location:	African origin, volunteers in waste or disturbed areas in South Florida.
Dormancy:	Evergreen
Culture:	Best in full sun, but stands light shade. Resistant to parched, dry conditions. Not cold tolerant below °F. This perennial is often grown as a ground cover using dwarf strains for dense cover, but not totally reliable because of fungus in wet weather.
Use:	Ground cover, border or facer, florists hanging baskets.
Other:	"Bright Eyes" dwarf white strain, Coquette dwarf rose.

Wedelia trilobata

WEDELIA

Wedelia: Latin for G. Wedel, German Botanist

trilobata: Latin for three lobed leaf

Size:	Up to 14 inches
Form:	Mat or carpet-like blanket of runners
Texture:	Medium
Leaf:	Variable from 3 lobed to simple rough toothed margins, dark green.
Flower:	Yellow, daisy-like 10-12 rayed, 1 to 1½ inch diameter with dark yellow disk.
Fruit:	Miniscule
Geographic Location:	West Indies to South America. South Florida coastal area. Introduced in late '20's to Palm Beach.
Dormancy:	Evergreen
Culture:	Salt tolerant behind primary dunes. Light sandy soils fairly draught tolerant. Full sun. Needs cutting back about once a year. If a high mower is used, apply fertilizer after.
Use:	As a ground cover maintain a 'fire strip' where other shrubs are contiguous. In planters atop walls, Wedelia cascades downward 10-20 feet. As a blanket it presents best appearance at 5-7 inches high.
Other:	Several dwarf forms are popular.

Zebrina pendula

WANDERING JEW

Zebrina: With Zebra stripes

pendula: Latin for pendulous or
hanging

Size:	10 inch mat
Form:	Carpet of intertwined succulent stems and leaves
Texture:	Medium
Leaf:	Succulent, ovate to oblong, reddish purple in sun or purple in shade. Silvery above with purple band down the middle and around the sides.
Flower:	Small, lavender, nested within a protective leaf-like bracket.
Fruit:	Tiny, inconspicuous
Geographic Location:	Tropical America. Coastal South Florida to Key West.
Dormant:	Evergreen. Cold tolerant to 28°F. for a short time.
Culture:	Best located in semi-shade for most of the day, well drained forest loam with adequate moisture. Notably free of pests. Plant: 6 inch cuttings on 6 inch centers, water regularly.
Use:	Besides being a colorful low cover, this plant has the ability to "eat up" fallen leaves, even large foliage like Seagrape for example. A word of caution: the brittle stems are easily destroyed by dogs and people walking through the bed.

220

VINES

The tropical regions of the Earth contain hundreds of species of vines, many of which are too tender to take up space in this book. Thus, the selections we have featured contain a group of vines normally found in gardens in the warmer parts of Florida.

Again, we stress the fact that a few of these plants are included not because they are scarce and hard to find, but for those people who are willing to search through nurseries ending with the great treat it can be, when you bring into flower a vine not known to your neighbors.

It is always good to extract information as to the vine's cold hardiness, its susceptibility to certain diseases, as well as other factors of growth. Certain planning before purchasing is necessary with vines: it may be in order to construct an arbor or trellis or maybe just a lattice work for vine support. Judicial selection of location based on space, sun and aesthetic suitability are considerations.

Vines differ from many other plants in that they should and will be selected for blooming characteristics more than texture, screening qualities or adaptability to wind or cold.

The writer learned a lesson in color coordination when he planted three handsome red Bougainvillea "Barbara Karsts" in an espalier arrangement on a wall behind a bed of specimen red Fireball Poinsettias. The color clash was heard all over the Peninsula of Florida!

Allamanda cathartica
'Hendersoni'

BROWN BUD ALLAMANDA

Allamanda: for F. Allamand,
a Dutch professor

Size:	Shrub or vine, depending on training, 4-5 feet x 4-5 feet.
Form:	Loose and spreading.
Texture:	Medium
Leaf:	Elliptic, to 6 inches long in groups of 3 or 4 , glossy light green.
Flower:	Bell-shaped, 5-lobed, 5 inches wide, golden yellow. Buds are brown.
Fruit:	Rare.
Geographic Location:	Brazil. Grows in South and Central Florida, subject to damage in cold weather.
Dormant:	Evergreen
Culture:	Best flowering in sunny locations. Fertilize 3 times per summer. Develops a leggy habit in several years. Judicious pruning is then recommended. Milky-white sap is very poisonous. Prefers rather dry moisture conditions.
Use:	Variety Hendersoni outdoes its parent A. cathartica, in all respects. It is usually fanned out on a fence or espaliered on brick and stucco walls. Choice of soils is optional.

Antigonon leptopus

CORAL VINE

Antigonon: Greek, referring to
   jointed flower stem

leptopus: slender stalked

Size:	Variable (can be a tree-top climber)
Form:	Dense climber with a pendant habit.
Texture:	Medium to coarse.
Leaf:	Heart-shaped, alternate, net-veined 3-5 inches long.
Flower:	½ inch long, five petals with sprays ending in branching tendrils. Masses of flowers, bright pink, draping walls or larger plants.
Fruit:	Not conspicuous. Seed case is small, triangular and brownish.
Geographic Location:	Mexico. North, Central and South Florida, subject to freezing conditions.
Dormant:	Evergreen in tropic areas only. Deciduous North of Palm Beach County.
Culture:	Sun-loving, not fussy on soils, tolerant to dry conditions. Recovers from ground-level if frozen. Propogates well from seed.
Use:	Climbs by tendrils. Bloom peak is late Summer and Fall. Its use is more adaptable to rural or farm-type sites than to urban properties since it volunteers where it can easily be classed as an escape.
Other:	There is a white form.

Beaumontia grandiflora

NEPAL TRUMPET FLOWER

Beaumontia: for Lady
    Beaumont of England

grandiflora: for large flower

Size:	Strong, woody climber
Form:	The vine will cover completely a pergola or old building. Requires strong supports.
Texture:	Coarse
Leaf:	Opposite, oval to obovate to 8 inches x 2 ¾ inches wide, simple entire, leathery, prominently veined beneath, wavy-edged.
Flower:	White, trumpet-shaped to 5 inches long and as broad; fragrant, greenish at base. Eye-catching.
Fruit:	Long, woody pods splitting lengthwise into 2 pods.
Geographic Location:	India, Nepal. In Florida in frost-free areas.
Dormant:	Evergreen
Culture:	For flower production use rich soil, sun and good drainage. After blooming it should be cut back severely to induce new growth.
Use:	Use on strong trellis in the garden, screens, on arbors, not to be used in small dooryard areas. Does well as a non-supporting mound in an open area.

Bougainvillea spectabilis

BOUGAINVILLEA

Bougainvillea: for M. de
    Bougainville, French
    navigator

spectabilis: visually striking

Size:	Intermediate
Form:	Upright sucker growth adds many feet per season.
Texture:	Medium
Leaf:	Dark green, ovate abruptly acuminate, 2 inches long.
Flower:	a). *B. spectabilis:* pubescent usually. Colors in the purple-red range, thorny stems, twigs and leaves are hairy. Leaves thicker and larger than b). *B. glabra:* smooth-leaved, less thorny species with rose-red flower bracts. Colors mostly clear as Barbara Karst and Afterglow.
Fruit:	Rare
Geographic Location:	Brazil. Central and South Florida. Freezing temperatures melt them to the ground. Will grow again in Spring.
Dormant:	Evergreen
Culture:	Cultivation in any soil, thrives best in full sun. Propagated by cuttings. Light applications of fertilizer three times per growing season. Normal watering. Leaf-chewing worms eliminated with liquid Sevin.
Use:	B. purple, as a rule, is easily trained into growing without support into a large shrubby mass. They can climb high in upright trees like slash pine and melaleucas. All are well-suited for espalier work. Astounding color accents can be found in potted specimens for pools or patios.
Other:	B. spectabilis: purple bracts. B. spectabilis sanderiana: red bracts. Barbara Karst, Crimson Lake, Afterglow, Apricot.

Clytostoma callistegioides

VIOLET TRUMPET VINE

Clytostoma: Greek for splendid mouth

callistegioides: callistegia-like (Bindweed)

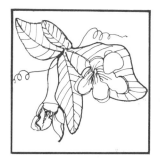

Size:	9 to 12 feet
Form:	Rank climber, tendril-supported.
Texture:	Medium
Leaf:	Leaflets; one or two elliptic oblong, to 4 inches long, undulate, glabrous, opposite leaves.
Flower:	To 3 inches long and wide at mouth, color lavender and streaked violet. Flowers axillary trumpet.
Fruit:	A broad prickly cap.
Geographic Location:	Bolivia, Brazil, Argentina.
Dormant:	Evergreen. Hardy to 20°F. Subtropic to temperate in Florida and southern Georgia.
Culture:	Full sun or shade, average water. Prune in Winter to control growth. Otherwise remove unwanted runners and old flower spikes. Tendrils will attach to open-type fences, trellises and arbors.
Use:	Not for use on small properties. Excellent cover for shade on pergolas and as a mix with other vines on major fences, both for wood and chain-link.

Dipladenia flava

YELLOW MANDEVILLEA

Dipladenia: double and ten

flava: yellow-colored

Size:	25-30 feet.
Form:	Rank vertical climber, rather dense, twining tip growth.
Texture:	Medium
Leaf:	Opposite, oblanceolate to oblong, 2 to 2½ inches long with recurved edges, midrib and veins depressed.
Flower:	Butter-yellow, showy tubular trumpet with 5 petals joined. Opening to 1¾ to 2 inches.
Fruit:	Not seen in Florida.
Geographic Location:	Columbia, Northern South America. Used occasionally in subtropic Florida.
Dormant:	Evergreen. Tolerant to marginal freezes only.
Culture:	Enjoys full sun with plenty of moisture, well-drained soils. Normal fertilizer, but do not overfeed. Loamy soils. May develop root-knot. Pests: orange oleander caterpillar, eradicate with Sevin.
Use:	This floriferous vine, blooming all Summer, is at its best when contrasted with purple bougainvillea on a trellis or pergola in alternate color notes.

Ficus pumila

CREEPING FIG

Ficus: ancient Latin for the fig

pumila: for dwarf,
        short-growing

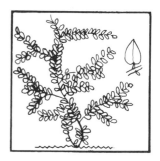

Size:	Can climb walls of a 4-story building.
Form:	When forming a vine, growth is ladder-like. Later, dense covering but loose.
Texture:	Fine to medium
Leaf:	Dark green. Leaves form two rows, are oval, blunt, 1 to 4 inches long. Clings to walls by means of a rubbery exudation from the roots.
Flower:	Inconspicuous
Fruit:	Green figs, the size of hens' eggs.
Geographic Location:	South China, Malaysia.
Dormant:	Evergreen
Culture:	Most soils, sun or shade, positive drainage, average moisture until well established. Probably should be considered high maintenance because of its voracious growth habit which should be pruned back in time.
Use:	The perfect wall-cover for energy conservation of air-conditioning. Also started with small (gallons) plants which will soon make a handsome tracery of creeping stems on walls. A year or so will fill the tracery into solid cover. A rapid grower. Periodic clipping is necessary.
Other:	F. minima: a midget of the genus

Hoya carnosa

WAX PLANT

Hoya: for T. Hoy,
   English gardener

carnosa: fleshy

Size:	Variable, but up to 10 feet.
Form:	Not distinguishable.
Texture:	Medium
Leaf:	Oval 2-4 inches long, medium green sometimes variegated in varieties. Leaves are thick and waxy.
Flower:	Tight round clusters of creamy white flowers, ½ inch diameter, each flower with a perfect 5-pointed pink star in the center.
Fruit:	Small pods if present.
Geographic Lcoation:	China and Australia. Subtropic Florida only. Super-sensitive to freezing.
Dormant:	Allow plant to become dormant during cool months by reducing waterings. Encourage foliage by Spring waterings.
Culture:	Soil must be nematode-free, such as peat moss, sphagnum or some fibrous, sterilized composts. Light or high shade. Propagate by cuttings and air-layering. Pests: mealy bugs and nematodes.
Use:	In Florida these plants are usually grown in containers: pots, boxes, tubs or hanging baskets to insulate roots from nematodes. Some plants are planted in the "boots" of date palms or sabal palms.
Other:	H. variegata: pinkish-white and green.

Lonicera japonica

JAPANESE HONEYSUCKLE

Lonicera: for A. Lonicer,
    German botanist

japonica: native to Japan

Size:	From semi-prostrate to climbing 10 to 15 feet.
Form:	Tends to mound on itself or also can form a blanket on slopes. A twiner.
Texture:	Medium
Leaf:	Opposite, dark green, leaf base obtuse to acute at leaf tip, 2 inches long by ½ inch wide.
Flower:	Over 1 inch long, tubular but quite narrow until petals open then recurve strongly. White, fragrant.
Fruit:	Black berries in clusters around stem in the Fall.
Geographic Location:	Japan, China and Korea. Eastern seaboard into the State of Florida.
Dormant:	Evergreen in Florida.
Culture:	Sun or shade. Great adjustability to soils. Roots itself when tendrils touch moist soil. Pest-resistant. Requires maintenance if allowed to grow rampant.
Use:	Its twining habit on chain-link fences produces an almost instant foliage screen. L. 'Halliana' is a popular trellis subject by virtue of yellow, fragrant flowers. Excells on slopes for erosion control.

230

Lonicera sempervirens

TRUMPET HONEYSUCKLE

Lonicera: for early German naturalist, A. Lonicer

sempervirens: Latin for evergreen

Size:	Variable
Form:	Twining vine
Texture:	Medium
Leaf:	Terminal leaves are joined together at their bases, perfoliate, most are simple, opposite, ovate and roundly pointed, 2 inches long.
Flower:	Long, slender, tubular, 1½-2 inches long, ending on five miniature petals recurved, in groups of 5, terminally orange-scarlet with orange throat. Winter and Spring.
Fruit:	Red berries in the Fall.
Geographic Location:	Generally South-east U.S. from Connecticut to Florida to Texas. Native. Cold-hardy.
Dormant:	Evergreen in Southeast coastal Florida.
Culture:	Tolerant of soil conditions, sunny exposure or light shade. Fast and heavy grower. Good drainage. Pests: subject to plant lice.
Use:	Excellent subject for post and rail fences, as well as wire or chain-link fences and gives total coverage. L. sempervirens also will soften old pines when planted at the base. One of our really colorful natives. Fall berries attract the birds. Trellises.
Other:	L. sempervirens var. sulphurea: a native yellow colorbreak of the above. Handsome.

231

Monstera deliciosa

MONSTERA OR CERIMAN

Monstera: a plant having an
abnormality in form

deliciosa: refers to the fruit
which flavors ice cream.

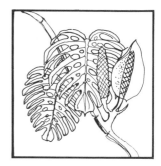

Size:	Variable
Form:	A giant aroid with 3-foot long leaves deeply and pinnately cut with oval perforations in the leathery leaf.
Texture:	Very coarse.
Leaf:	See "Form", also close-jointed tree climber forming long hanging cord-like aerial roots.
Flower:	Bisexual spadix clasped by a large, fragrant, creamy white spathe.
Fruit:	Rich, orange-red kernels, soft when ripe and edible, aromatic, issuing from an "ear of corn" or spadix.
Geographic Location:	Mexico and Guatemala. Pan tropic coastal South Florida
Dormant:	Evergreen. No frost-tolerance.
Culture:	Rich forest loam. As a juvenile it is a fast growing tropic liana needing the support of tree trunks with rough bark. It must have light shade in Florida. Old plants are slow-growing. Young plants of monstera deliciosa are often offered as philodendron pertussin.
Use:	Best purchased as vining plants in pots or as established mature leafed specimens (4 to 6 leaves) for pot or patio use. One of the oldest and most popular of rich coarse-leaved tropical serving many purposes.

Monstera friedrichstahli

SWISS CHEESE VINE

Monstera: a plant having an
abnormality in form

friedrichstahli: for a
German naturalist

Size:	Variable
Form:	Multi-stemmed, tropical tree climber.
Texture:	Coarse
Leaf:	12-14 x 7½ inches, tough, ovate, wavy margined. Leaf with an acute tip, obtuse base perforated with elliptic holes randomly on either side of the midrib. Petiole 13-15 inches long, upper side is channeled, nodes 2½-3 inches apart.
Flower:	Spathe is creamy-white, 6-8 inches long.
Fruit:	Spadix 5 to 6 inches containing kernels like "grains of corn" on a cob.
Geographic Location:	Central America, Caribbean including Trinidad. Pan tropic South Florida, South of frost line.
Dormant:	Evergreen
Culture:	Given shade and moisture, this subject will produce multi-tips from a single plant on a sabal palm. Any soil composition, lots of organic fertilizer. Natural in rain forest conditions. Pests: none.
Use:	M. friedrichstahli probably should be used more than its specie counterpart because of ease in climbing and its vigorous make-up. It can climb wood fences, rubble walls and would blend nicely with epipremnum aurea.

Pandorea jasminoides

BOWER VINE

Pandorea: Greek for beautiful leaf

jasminoides: similar to jasmine

Size:	20-30 feet, slender twisting stems.
Form:	Vining
Texture:	Medium to fine.
Leaf:	Glossy foliage, medium to dark green. Leaves have 5 to 9 egg-shaped leaflets 1 to 2 inches long. No tendrils.
Flower:	White with pink throat, 2 inches long, in small clusters, funnel-shaped with 5 lobes, spreading.
Fruit:	Capsules 3 inches long, oblong and stiff-valved.
Geographic Location:	Australia. Southeast coastal Florida. Gulf coast North to Sarasota.
Dormant:	Evergreen. Withstands a little frost for a short time.
Culture:	Rich soil, sunny location. Best to protect from strong winds. Good drainage, normal fertilizer in growing season. Pests: some nematodes, scales, caterpillars occasionally.
Use:	Trellis for this vine should be 10 feet x 5 feet, minimum. Can be used on 8-foot chain-link fences with occasional pruning control. Best used in wind-free patio areas.
Other:	P. alba: has white flowers. P. rosea: has pink flowers.

Passiflora edulis

PASSION FLOWER,
   PURPLE GRANADILLA

Passiflora: for flower park
   resembling the Holy
   Trinity, the Crown
   of Thorn

edulis: Latin for edible,
   referring to fruit

Size:	Climbing 20 to 30 feet by tendrils.
Form:	Vigorous and prone to tangling.
Texture:	Medium
Leaf:	3-lobed leaf deeply toothed, to 3 inches long, leathery. Alternate.
Flower:	White with white and purple crown, 2 inches at leaf axils.
Fruit:	Deep purple, fragrant, 3 inches long, in Spring and Fall.
Geographic Location:	Brazil. Southeast coastal Florida.
Dormant:	Evergreen
Culture:	Full sun, average watering and fertilizer, well-drained soil. Protect from strong winds. Fairly pest-free.
Use:	On trellises or walls and fences for their vigor and bright showy flowers. Also for bank-holding properties.

Petrea volubilis

QUEEN'S WREATH

Petrea: for Baron Petre,
    English patron of botany

volubilis: climbing by embrac-
    ing another object

Size:	Approximately 25-30 feet.
Form:	Twining, woody vine.
Texture:	Coarse, very rough.
Leaf:	Pointed oval to elliptic, 4-8 inches long, petiole 2 inches long. Rough sand-papery upper and lower sides. Opposite.
Flower:	Showy racemes of lovely star-like flowers of long lilac-blue sepals and small violet carolla 1½ inch in diameter.
Fruit:	Drupes which include long-lasting calyces which serve as wings for the single seed.
Geographic Location:	Mexico, Central America, West Indies. Coastal South Florida.
Dormant:	Evergreen. No cold-hardiness. Protect from frost when imminent.
Culture:	Prefers a rich, sandy soil, but tolerates other soils. Good surface drainage. Sunny exposure. Average fertilizer, 6-6-6 in Florida with trace elements. No apparent pests.
Use:	One of the rare (in the tropics) blue-flowering vines, a bit similar to wisteria. Trains well for lattices, arbors and trellises. Ever popular.
Other:	P. volubilis albiflora has white flowers.

Philodendron hastatum

NO COMMON NAME

Philodendron: Latin for
  tree-loving

hastatum: Latin for having a
  triangular shape like
  an arrowhead

Size:	35 feet high x 4-6 feet spread
Form:	Climber, fairly rapid, by aerial roots.
Texture:	Coarse
Leaf:	36 x 18 inches, fresh green lush climber with mature leaves hastate and undulant.
Flower:	Gorgeous inflorescence with tubular pale green spathe, red inside.
Fruit:	Spadix, vertical spike with kernels as in an ear of corn.
Geographic Location:	Brazil, Tobago and Trinidad. Coastal South Florida.
Dormant:	Evergreen. Withstands 28°F for short periods of time.
Culture:	Rich sandy forest loam. Prefers some shade for ideal growth. Best growth is in moist, high humidity. Fertilize regularly. No serious pests.
Use:	Striking when planted at the base of sabal palms and rough-barked trees like live oaks. Use on rubble or brick walls as well as solid wood fences. Nature employs this plant as a natural cover on slopes and banks.

**Philodendron scandens,
subsp. oxycardium**

GRANDMOTHER'S
PHILODENDRON

Philodendron: Latin for
tree-lover

oxycardium: heart-shaped
with pointed tip

Size:	High-climber developing into many vertical stems in time.
Form:	When climbing upwards leaf sizes increase measurably. When descending on the same long stem leaf sizes will decrease to the juvenile stage.
Texture:	Coarse to medium depending on application.
Leaf:	Deep shining green, broadly heart-shaped. Small in juvenile stage to 12 inches long in maturity.
Flower:	Typical aroid flower, cream-colored.
Fruit:	Vertical spike or cylinder with seeds protruding when ripe.
Geographic Location:	Puerto Rico, Jamaica and Central America. Coastal South Florida.
Dormant:	Evergreen. Tolerant of freezing conditions to 28 or 30°F for short periods.
Culture:	Rich, sandy forest loam. Dappled shade to full shade. Heavy on moisture, good drainage, two applications of fertilizer per growing season. No apparent pests.
Use:	Deserves notice for its ivy-like quality when it cascades down from raised boxes or planters. Also at the base of coarse-barked trees, walls or fences for climbing. See "Form" for note on leaf habits on most philodendrons.

Philodendron radiatum

CUT-LEAF CLIMBING
PHILODENDRON

philodendron: tree-loving

radiatum: Latin for "in
radiating manner"

Size:	25 to 35 feet with multiple stems.
Form:	A strong climber whose cut-leaves give this plant a medium texture.
Leaf:	A lush-climber with broad, rich green leaves deeply lobed; 18-24 inches long, in juvenile stage less incised and often known as P. dubia, commercially.
Flower:	Typical aroid spathe and spandix
Fruit:	Typical vertical column which, when ripe, exudes seeds.
Geographic Location:	Southern Mexico and Guatemala. Coastal South Florida.
Dormant:	Evergreen. Protect from freezing conditions.
Culture:	Rich forest loam, sandy texture, shade to dappled shade. Generous growth on moisture and well-drained soil. Fertilize 2 times per growing season. No apparent pests.
Use:	A more delicate climber than P. hastatum or P. giganteum. Often used at base of trees like coconut or sabal palms, or live oak trees. Also performs well in large concrete architectural planters. Will do well on shady rubble walls and fences.

Podranea ricasoliana

PODRANEA VINE

Podranea: a word made for
    another; transposed
    letters, an anacronym

ricasoliana: from Italian
    gardens of Ricasoli

Size:	Moderate height climber
Form:	Scrambles, climbs with support and ties.
Texture:	Medium
Leaf:	Evergreen, odd-pinnate, often 9 leaflets, no tendrils, pointed, up to 2½ inches each. Semi-lacy effect.
Flower:	Bell-shaped, pale pink with red penciling interior of tube, 2 inches long in hanging terminal panicles.
Fruit:	A slender pod up to 12 inches long which splits into 2 parts.
Geographic Location:	South Africa. Frost, even lightly exposed, is likely to kill everything but roots.
Dormant:	None
Culture:	Requires ample sunlight to light shade. Fertile soil. Water frequently and train on a support. Sometimes attacked by nematodes.
Use:	Although an old standby, its lacy quality and delicate coloration make it undiminished in popularity among vine lovers. Trained on fences or any vertical lattice-work such as a porch or arbor, it is one of our best vines.

240

Pyrostegia ignea

FLAME VINE

Pyrostegia: Greek for fire on a shelter

ignea: Latin for flame-colored

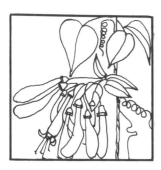

Size:	20-30 feet in trees and telephone poles.
Form:	Vine, climbing by use of tendrils.
Texture:	Medium
Leaf:	Leaves have 2 or 3 leaflets, and a 3-parted tendril. Each leaflet ovate, 2-3 inches long.
Flower:	Deep orange in abundance in January and February, in pendant clusters 2-3 inches long with 5 recurving points. Corollas is a long narrow tube.
Fruit:	Not apparent in Florida. 12-inch capsule in Brazil.
Geographic Location:	Brazil. Coastal South Florida.
Dormant:	Evergreen. Damaged by 32°F temperatures.
Culture:	Soils variable with average fertility added occasionally. Positive drainage. Sunny exposure for maximum color in blooms. No pests.
Use:	Traditionally used for smashing orange inflorescence in January. Plant for excellent cover on arbors and trellises as well as wood fences and lattice-work. Will take over trees if this is allowed.

Senecio confusus

MEXICAN FLAME VINE

Senecio: Latin for old man

confusus: Latin for uncertain
or confused

Size:	15 to 20 feet x 3 to 5 feet.
Form:	Vining and tangled.
Texture:	Medium
Leaf:	Tongue-shaped with coarsely toothed serrations to 4 inches long. Alternate arrangement.
Flower:	Daisy-like rays and center of 1½ inch diameter, vivid orange, deepening to red-orange in clusters. Blooms in Spring and Summer.
Fruit:	Little, bristled achenes with filamented attachment for wind distribution.
Geographic Location:	Mexico. Coastal South Florida. Cold damage under 25 to 28°F.
Dormant:	Evergreen
Culture:	Full sun or shade. Tolerant of soil variations. Once established senecio will thrive with little attention. Some pruning is needed. Pests: caterpillars, scales and mites.
Use:	This self-supporting vine will creep over fences, up trees and around rural mailbox supports.

Solanum wendlandii

COSTA RICA NIGHTSHADE

Solanum: referring to
sedative affects

wendlandii: man's name

Size:	Variable
Form:	Vigorous woody vine with spiny stems which twine, few thorns.
Texture:	Medium
Leaf:	Two types; simple or deeply 3-lobed to 10 inches for the latter. Some prickles on midrib and leaf stems.
Flower:	Lavender-blue to 2 inches in diameter. In large trusses 1½ feet across, terminal heads. Round.
Fruit:	Inconspicuous
Geographic Location:	Costa Rica. Best grown in coastal South Florida without frost.
Dormant:	Evergreen, except is defoliated in a freeze. Slow to leaf out in Spring.
Culture:	Ordinary garden care, average soils, good drainage. Sun to dappled shade.
Use:	Will clamber over trees, best used on fine to medium textured trees. Covers pergolas and decorates overhanging eaves of some large houses.

Stephonotis floribunda

BRIDAL BOUQUET,
   MADAGASCAR
   JASMINE

Stephanotis: referring to
   white crown

floribunda: Latin for
   profusely flowering

Size:	15 to 20 feet.
Form:	Twining vine, woody stems.
Texture:	Medium
Leaf:	Oval, leathery, shiny, opposite, smooth-edged, up to 4 inches long. Fast-growing tendrils.
Flower:	Funnel-shaped, white, waxy, fragrant in leaf axils, 1-2 inches long in open clusters. 5-pointed and flared petals at lip of flower.
Fruit:	A horn-like seed pod to 4 inches long.
Geographic Location:	Madagascar and Malaya. Coastal South Florida.
Dormant:	Evergreen
Culture:	Moderate growth. Roots do best in shade with dappled sunlight for vines. Liberal moisture. Rich, sandy loam with positive drainage. Pests: scale, mealy bugs, nematodes.
Use	Favorite for bridal bouquets. Use on a trellis but in close relation to entrance door for full impact of flowers and fragrance. A neat vine.

Stygmaphyllon littorale

ARGENTINE AMAZON VINE

Stigmaphyllon: Greek,
combining stigma
and leaf

littorale: oriented to the beach
or waterfront

Size:	Vigorous rank climber, twining to the tops of tall trees if allowed, 20-30 feet or more.
Form:	Semi-compact to loose irregular.
Texture:	Medium
Leaf:	Opposite, simple ovate, 3½ to 5 inches long.
Flower:	Clusters of malpighia-type flowers 10 to 20 in clusters, bright yellow 1 inch across. Sometimes likened to Oncidium blossom.
Fruit:	Winged fruit.
Geographic Location:	Argentina. Central (near coast) and Southern Florida.
Dormant:	Evergreen
Culture:	Rich soil, ample water. Requires a climbing support structure. Prefers shade or dappled sun. Prune out dead or straggling growth. Fast-growing. Tolerates some mild winter weather. Allow some protection.
Use:	Well-known for its summer flower color which is striking. Use on trellis-work, arbors, pergolas and tall chain-link fences.

Tecomaria capensis

CAPE HONEYSUCKLE

Tecomaria: from Tecoma, and
old genus name for
Bignonia family

capensis: referring to the
Cape of Good
Hope, Africa

Size:	8 feet to 20 feet.
Form:	Scrambling vine, but with hard pruning can become an upright shrub.
Texture:	Medium to fine.
Leaf:	Soft feathery foliage, compound, 5-9 odd pinnate, serrated leaflets. Somewhat fern-like.
Flower:	Brilliant orange-red, tubular, 2 inches long, slightly flaring into 4 segments in elongated clusters.
Fruit:	Narrow, flattened capsules are from 2-7 inches long, freely produced.
Geographic Location:	South Africa. Coastal South Florida. Temperature tolerance to possible 28°F.
Dormant:	Evergreen
Culture:	Good garden sandy loam. Sunny exposure. Positive drainage. Welcomes summer fertilizing. Average moisture. Adapts well to pruning and shearing. Pests: scales, mites.
Use:	A popular subject when espaliered. Effective as a vine on a carriage post light or on a low post and rail paddock fence. Has been used as a ground cover on banks and slopes.

Thunbergia fragrans

White Thunbergia

Thunbergia: after Peter
      Thunberg, 1743-1828

fragrans: fragrant

Size:	Vigorous, mounding, 6 to 8 feet.
Form:	Woody twiner for trees and fences.
Texture:	Medium
Leaf:	Opposite, lanceolate to triangular ovate, to 3 inches long, nearly entire, finely toothed, medium soft green.
Flower:	Solitary, corolla snow white, 1¼ inch long x 2 inches across, singly or clustered, sometimes fragrant.
Fruit:	Rounded seed capsule ends in a beak.
Geographic Location:	India and Sri Lanka. South Florida only.
Dormant:	Evergreen, tropics only.
Culture:	Rich, sterilized top-soil. Bright sun to high shade. Generous waterings and good drainage. Protect from strong winds. Pests: none other than nematodes.
Use:	For use on trellises, porches and arbors. The multitudes of white flowers are a happy departure from the quantities of gay-colored tropical vines we live with. Tends to run wild without maintenance.

Thunbergia grandiflora

SKY FLOWER OR
BENGAL CLOCK VINE

Thunbergia: after
Thunberg, botanist

grandiflora: Latin for
large-flowered

Size:	To 20 feet.
Form:	Vigorous twiner.
Texture:	Medium
Leaf:	Broad, heart-shaped, angularly lobed, 3-8 inches long, pointed tip.
Flower:	Lavender-blue, bell-like with a white throat, somewhat pendant, two-lipped in leaf axils.
Fruit:	Seed capsule to ¾ inch long, beaked.
Geographic Location:	India and world tropics. Coastal South Florida.
Dormant:	Evergreen. Tropical climate only.
Culture:	Best in rich, sterilized top-soil. Sun to partial shade. Generous watering and good drainage. Protect from strong winds. Pests: none other than root-knot.
Use:	Its cascading flower effect calls for a rather high wire fence or lattice-work. A handsome presentation can be made by training a vine on a 2-story building.

Trachelospermum jasminoides

CONFEDERATE JASMINE

Trachelospermum: Greek combined for neck and seed

jasminoides: jasmine-like

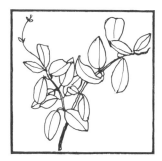

Size:	To 20 feet when supported.
Form:	Twisting vine, woody and climbing, with long branch tips which are usually tipped back in Spring.
Texture:	Medium-fine.
Leaf:	Ovate-lanceolate to 2½ inches long, opposite, dark green, glossy above, lighter below, prominent midrib.
Flower:	Pure white, 5-petaled pinwheel, counterclockwise, 1 inch spread, fragrant, on long-stemmed petioles in leaf axils.
Fruit:	The twinned pod-like fruit slightly spreading.
Geographic Location:	China
Dormant:	Evergreen. Fairly hardy to low temperatures even in North Florida.
Culture:	Slow to start, but eventually moderately fast. Be sure to stake this vine at the time of planting. Soils and watering are average. Pests: scale, insects and sooty mold.
Use:	Train in espalier fashion, on post lights, arbors, lattice-work or trellises, also mailbox posts. Sometimes employed as a ground cover, then cut out vertical shoots.
Other:	T. jasminoides minima: a dwarf ground cover.

ACCENT PLANTS

The name Accent Plant implies a group of mostly tropical species that are identifiable by their extremes of texture such as Bananas (of coarse texture) and Fountain Grass (of fine texture). The palette of temperate-zone plant materials, by comparison, contains very few of these types of plants.

Accent Plants are used as exclamation points at important locations in the planting design where they act as sculptural accents, as given a group in a bed of low Ground Covers. These plants also can be arranged at uniform intervals to tie together plant groupings in a linear design. They also can be the dominant plants in a plant combination around a swimming pool, atrium or patio where these textures are similar to those of the world's tropics.

Accent Plants cover a broad field of textures as well as sizes. When used within the bedlines of Ground Covers, the sizes of accent plants will be comparatively small, but larger than the surrounding Ground Covers. Then again, the Accent Plant may even be a tree such as character types found in some irregular Podocarpus specimens, or the Bahama Spiney Bucida or Pandanus Tree.

Aglaonema commutatum

AGLAONEMA

Aglaonema: Greek for bright
 thread, referring
 to stamen

commutatum: referring to
 changes in leaf color

Size:	2½-3 feet x 2 feet.
Form:	Vertical stem unbranched. Large alternate leaves growing out in all directions.
Texture:	Medium
Leaf:	Ovate, pointed, leathery leaves, medium grayish green, largely variegated with silver, 9 inches long, 5 inches wide (petiole 5 inches).
Flower:	Resembles miniature Callas, white. The spathe drops off.
Fruit:	Olive-shaped, cardinal red, 3 months to germination. Variegation does not reoccur from seed.
Geographic Location:	Malay and Borneo. Tropic Florida only.
Dormant:	Evergreen
Culture:	Low level of light if needed, well-drained, organic soil. Admix of peat and perlite admissable. Continuous moisture. Pests: subject to nematodes, mites and scales.
Use:	Other than its successful use in pots and planters indoors, its mottled foliage of silvers and cool greens when introduced in colonies at the edge of quiet pools is non-pareil, but always in deep shade. Face it down with Peperomia.

Alocasia macrorhiza

GIANT ELEPANT EAR

Alocasia: anagram of colocasia by twisting letters

macrorhiza: large under-
ground rhizome

Size:	15 feet, but half that size in Florida.
Form:	Huge upright with bold shieldlike leaves in triangular shape.
Texture:	Coarse.
Leaf:	Broadly arrow-shaped, fleshy leaves, waxy green, prominent veins, wavy margin. 4 x 2½ feet spread of leaf with stem 4½ feet. Leaf stem green with brown speckles.
Flower:	Tiny flower on spike surrounded by greenish white bract.
Fruit:	Reddish fruits form on spikes much as corn on the cob.
Geographic Location:	Malaya, Sri Lanka.
Dormant:	Evergreen
Culture:	Soil fertile with organics, lots of water, high shade. Always needs wind protection. Subject to freeze damage. Propagated by divisions of underground stems.
Use:	Used in a colony much like bananas to strengthen the feeling of the tropics. Protected moist areas such as patios, shaded understory locations, and ravines.

Alpinia zerumbet

SHELL GINGER LILY

Alpinia: for P. Alpinus, early
　　Italian botanist

zerumbet: old East
　　Indian word

Size:	8-9 feet x 6-8 feet.
Form:	Heavy rosette of stems and leaves symmetrically arching out from a large loose center. Suckers.
Texture:	Coarse
Leaf:	Double row of leaves 2 feet long x 5 inches wide, waxy dark green, ginger odor when crushed. Each cane is replaced once a year.
Flower:	Pendant clusters appear terminally on arching canes every Spring. Waxy white or pink. Shell-like.
Fruit:	Rare
Geographic Location:	Polynesia. Above ground survival in South Florida. Roots survive to about 15°F for Central and North Florida.
Dormant:	Evergreen
Culture:	Fertile, moist soil, partial shade to sunny. Fertilize 1 to 2 times per year. Prune out last year's flowered canes. Protect from strong winter winds. Thrives on much water.
Use:	This ginger will require at least double to triple its original spread in five years. Use in sun traps such as right angle corners in fences or buildings. Also effective around ponds and lakes. Unequaled for use in a tropical setting.
Other:	A. zerumbet variegata: semi-dwarf and spectacular.

Alpinia zerumbet variegata

VARIEGATED SHELL
GINGER LILY

Alpinia: for P. Alpinus, early
Italian botanist

zerumbet: old East
Indian word

variegata: marked in different
colors in diagonal, varied
width bands

Size:	Not over 3½-4 feet unless staked. More spreading or semi-procumbent.
Form:	Loose, rather broad, irregular.
Texture:	Coarse
Leaf:	6 to 12 alternate leaves per cane, height of cane to 3½ feet. Leaves 12 to 16 inches x 5 inches wide, elliptic lanceolate. Green and ivory striations diagonally up and out from midrib. Color bands irregular from pencilling to wide bands. Variegations are not symmetrical.
Flower:	Not seen in Florida.
Fruit:	Not seen in Florida.
Geographic Location:	South Pacific Islands. Pan tropic Florida.
Dormant:	None. Some tolerance to light freezes, roots hardy to 15°F.
Culture:	Prefers light shade to grow to maximum size. With sun exposure height is more horizontal, canes not vertical. Must be wet to moist. Fertilize three times in Summer. Remove year-old canes. No pests of note.
Use:	This semi-dwarf plant is not only an accent, but also doubles as a smashing ground-cover where it blends well with Spathiphyllum, Crinum, Bird of Paradise and Hawaiian Ti. As a ground cover use 2 or 3 rows with 3-foot spread plants on 4-foot centers.

255

Anthurium huegelii

NO COMMON NAME

Anthurium: Greek for flower
and tail

huegelii: for Carl Hugel,
German traveller.

Size:	3½ feet x 3½-4 feet.
Form:	Rosette with leaves angled from 11:00 to 3:30.
Texture:	Coarse
Leaf:	3 feet long x 10-inch spread. Leathery. Wavy margined, tapered tip. Midrib and lateral veins prominent. Usually dark green, coarsely shiny.
Flower:	Spadix 5-6 inches long x ¾ inch diameter when seeding, brown on long stem issuing from base of rosette 2 feet high.
Fruit	Bright red seeds, rice-size, issue from spadix.
Geographic Location:	Peru, Andes rain forests, Caribbean, Southern Florida.
Dormant:	Evergreen
Culture:	High fertility, good drainage, high moisture, strictly tropical with best growth at 80-90°F, but can withstand 32°F without damage. Protect foliage from sun and strong winds. Pests: slugs and snails.
Use:	Accent plant extraordinary where anthurium leaves can be silhouetted but always with a canopy shade. Great tub plant on paved decks for patios and atrium. Usually is ideal around pools and ponds. This plant can be arboreal in its native jungles.
Other:	A. hookeri, A. cubense, both equal in form to above.

Asparagus macowanii

REGAL FERN

Asparagus: a winter bud
    forming a scaly shoot

macowanii: for Peter Mac
    Owan,   1830-1909,
    English

Size:	6 feet x 6 feet.
Form:	Fountain-shaped to erect shrub. Foliage like billows of green smoke, extra fine texture.
Texture:	Fine
Leaf:	Stiff but threadlike, ¼ inch long. Light bright green on many long, slender stems arise in clumps from tuberous roots.
Flower:	In racemes to 1 inch long, white.
Fruit:	Berries scarlet, small. Rare in Florida.
Geographic Location:	Natal, South Africa. Tolerant to 22°F in Florida.
Dormant:	Evergreen
Culture:	Good drainage, fertile loamy sand. Dappled or high shade, keep moist. Fairly rapid grower. Propagated by tuberous root divisions. Note: formerly A. myriocloides.
Use:	As an accent in high shade area. Use with a ground cover of Juniperus procumbens compacta, Cyrtomium falcatum or Ophiopogon japonicus. Also is an attractive feature as a hanging basket, as a pot plant. Rates a 10 in my book for Asparagus.

Asplenium nidus

BIRD'S NEST FERN

Asplenium: Latin name for
    this fern

nidus: a nestlike structure
    bearing spores

Size:	4 feet x 6 feet
Form:	Upright cluster, rosette-shaped.
Texture:	Coarse
Leaf:	4 feet long x 8 inches wide, apple green, stiff erect in a large rosette, entire, wavy margin. Black dominant midrib.
Flower:	None
Fruit:	Spores
Geographic Location:	Polynesia and Asia, tropical South Florida to East Stuart.
Dormancy:	Evergreen
Culture:	This fern is epiphytic and terrestrial and can be mounted on a shady limb using copper wire. To establish in a bed use a fibrous loose high humus soil with extra watering to establish. Pests: snails and slugs. Propagated by spores.
Use:	When used in the garden this fern is a striking show piece and can be very successful as an accent especially using a group of 3 in staggered sizes. As a patio tubbed specimen it will be an attention getter. A classic form of fern.

Bambusa disticha

FERNLEAF BAMBOO

Bambusa: Latin version of
Malayan vernacular

disticha: producing leaves in
two opposite rows.

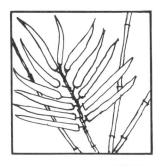

Size:	10 feet x 10 feet.
Form:	Slightly arching with many canes, dense clumps.
Texture:	Fine
Leaf:	Foliage alternate 1-6 inches long, flat. Light red.
Flower:	None
Fruit:	None
Geographic Location:	China, Japan. Hardy to 15°F. Most of Florida.
Dormant:	Evergreen
Culture:	A robust grower in rich soil with water and fertilizer. More conservative in growth under more spartan conditions. Well-drained soil, sunny exposure. Pests: not serious.
Use:	As a specimen standing alone, a lovely fountain shape. When planted as a thick, dense hedge it can stand shearing and thus develops a rigid impenetrable formal or linear shape. Its texture and color make this subject stand out and give its accenting style.

Bromelia pinguin

WILD PINEAPPLE

Bromelia: for Swedish
   botanist, M. Bromel

pinguin: with thick leaves

Size:	3-4 feet x 6-foot spread.
Form:	Rosette of stiff leaves, spiny.
Texture:	Coarse
Leaf:	6 feet long x 1½ inches wide, edged with spines curved forward and backward about ⅓ inch long. Can change color from green to red. Spines on leaf margins are lethal.
Flower:	Spectacular spikes of white meal. Reddish petals are 1 inch long, jointed nearly ¼ of their length. Tips are white, wooly.
Geographic Location:	West Indies, South America.
Dormant:	Evergreen
Culture:	Loose fibrous soil, well-drained. Occasionally needs watering. Feed lightly once or twice per Summer. Best color under sunny conditions. Rapid grower.
Use:	Ideal as a group of three, when arranged as a colony in a natural setting with a rotting snag of an oak tree trunk with ferns. Considerable maintenance is required to control the rank growth of this plant.

Cordyline terminalis

TI PLANT

Cordyline: referring to Greek
for club, to
thickened roots

terminalis: at the end of
the stem

Size:	8 feet x 3 feet. In Hawaii to 12 feet.
Form:	Single or multiple canes growing vertically with ± 10 leaves at the tip of each shoot, branched or unbranched.
Texture:	Coarse
Leaf:	Narrow, oblong leaves 1 to 2 feet long x 4 to 5 inches wide, arranged in a close spiral at the terminus of each shoot.
Flower:	Panicle 12 inches tall, flowers numerous, lilac-tinted, ⅓ inch long. Stem is ringed with leaf scars.
Fruit:	Small berries from yellow to red. Scarce.
Geographic Location:	Islands or Polynesia. Tropical, South Florida to 32°F.
Dormant:	Evergreen
Culture:	Soil well fortified with peat and cow manure. Good drainage. Regular waterings. Dappled to high shade. Fertilize three times per year. Protect from winds. Propagated by cuttings. Pests: Nematodes.
Use:	Leaves can be highly colored or plain dark green. In shrub groups it should be used in colonies mixing foliage colors overall, but keeping only one color per colony. Considered prime for use in stressing the tropical effect.
Other:	Ti is the strain with pure green foliage. Peter Buck: brought to Hawaii from Samoa by Sir Peter Buck, has orange-bronze leaf color; a color break in the family.

Cortaderia selloana

PAMPAS GRASS

Cortaderia: outer layer smoothed off

selloana: saddle-shaped

Size:	To 20 feet, usually 8-10 feet x equal spread.
Form:	Fountain of saw-edged grass, bearing above the grass 3 foot plumes of flowers, white.
Texture:	Medium
Leaf:	3 to 9 feet long x ¾ inch wide with rough edges, in thick clumps becoming woody at the roots.
Flower:	Beautiful white silky plumes to 3 feet, late Summer to Fall.
Fruit:	Unknown
Geographic Location:	Argentina, anywhere in Florida.
Dormant:	Is dormant where hard freezes occur. Comes back from the root.
Culture:	Female plants produce the well-known plumes. Will thrive in any soil. Fast-growing in rich soil in mild climates. Resistant to winds. Remove old bloom stalks at end of Fall.
Use:	Large bank plantings and for first line defense in a wide windbreak. It is nicely used as an underplanting for Sabal Palms. Do not use this plant in small-scale landscape work such as small residences. Best for large scale development.

Crinum amabile 'Purple Leaf'

QUEEN EMMA LILY

Crinum: Latin lily-like

amabile: obscure

Size:	6½ feet x 6½ feet, a giant among the crinums.
Form:	Leaves semi-rigid and radiating out from 6:00 to 4:00, similar to sisal agave in juvenile stage.
Texture:	Coarse
Leaf:	Broadly lance-shaped 5½-6 feet long x 4½-5 inches wide, occasionally the wind "dog-ears" the leaf. Green to deep blood-red.
Flower:	Has a red tube and is red down the center of the white segments which are 4-5 inches and ½ inch broad.
Fruit:	A bulblet, glossy deep red, shaped like a squat bottle with a prominent neck. Varies in size ¾ to 2 inches.
Geographic Location:	Sumatra, Hawaii and Southern Florida.
Dormant:	Evergreen
Culture:	Rather dry, well-drained soil, with a sunny exposure. Protect from strong winds. Tolerates some cold to 28°F. Mostly free of pests.
Use:	A form, color and accent plant. At least three plants set in a triangle on 6 to 8-foot centers for large scale projects. Once featured at Waikiki's Royal Hawaiian Hotel, a bed of 100 Queen Emma lilies was a memorable picture.

Crinum asiaticum

GRAND CRINUM LILY

Crinum: Latin, lily-like

asiaticum: from Asia

Size:	6 feet x 5 feet.
Form:	Big scale upright rosette of 18-30 broad green leaves taper pointed 3 feet long x 4 inches broad.
Texture:	Coarse
Leaf:	Leathery, thick, 3 feet long, dark green.
Flower:	White, spidery, 20 or more atop weighty stem, standing above the arching leaves.
Fruit:	A bulb variable in size, freely rooting even when not in soil.
Geographic Location:	Tropic Asia. Foliage damage below 31°F, but plants recover.
Dormant:	Evergreen
Culture:	Well-drained rather dry soil, sunny exposure, salt-tolerant. Protect from strong winds. Reaches its maximum height in about 5 to 6 years. Most often planted too densely.
Use:	One of the architect's favorite accent plants and one that blooms most of the year. Makes a fine underplanting for Seagrape trees and Sabal Palms. It blends well with the yuccas, and mixed with a bird of paradise specimen.
Other:	Crinum amabile: Queen Emma Lily, cerise flower.

Cycas revoluta

SAGO PALM

Cycas: Greek for palm tree

revoluta: Latin for rolled down
   at the tips.

Size:	To 10-12 feet, usually much smaller, 3 feet high x 3 feet spread.
Form:	Feather-like leaves grow out in rosette to 3-3½ feet in diameter.
Texture:	Fine
Leaf:	2 to 3 feet long divided into many narrow, leathery, dark green segments.
Flower:	Female: a flat cone. Male: a 20-inch cone with an offensive odor.
Fruit:	Red, somewhat flattened, pullet-sized eggs.
Geographic Location:	South Japan to Java. Hardy to 10°F. Central and South Florida.
Dormant:	Evergreen
Culture:	High soil tolerance. Shade or sun. Average water and drainage. Confined small root system. Pests: subject to scales and leaf-spot.
Use:	Traditionally a very happy pot subject maintaining its size for years. Effective accent plant in small areas such as bathroom gardens to produce a palm-like effect. Usually a single plant will do the job. Blends well as an accent colony with Liriope Evergreen Giant, both underplanting for a cluster of Sabal Palms.

Cyperus alternifolius

UMBRELLA PLANT

Cyperus: antique Greek name
for subject plants

alternifolius: alternate-leaved

Size:	2 to 6 feet high, triangular stems.
Form:	Upright sedge-type growth, sometimes wider than high in old plants. Symmetrical.
Texture:	Fine
Leaf:	The name "umbrella" describes the daisy-like wheel laying flat on its axis atop the reedlike stem, 20 rays, 8 to 16 inches in diameter.
Flower:	Within the axis occur the minute greenish flowers.
Fruit:	Indistinct and minute. Propagated by divisions and planting the "daisy wheel" in sand.
Geographic Location:	Madagascar, Phillipines and world-wide tropics.
Dormant:	Evergreen
Culture:	Not fussy as to soil, sun or shade. Dry soil as well as slightly submerged aquatic conditions. Controls spread by confining roots.
Use:	Wetland margins, bog gardens. Flanking foot-bridges at their heads. In pots and tubs or flanking natural waterfalls.

266

Cyperus papyrus

EGYPTIAN PAPYRUS

Cyperus: old Greek for
the plant

papyrus: paper-like, refers to
pulpy stem of cane

Size:	6-8 feet, eventually obtaining an almost equal spread.
Form:	Large, handsome sedge, strong, vertical canes.
Texture:	Fine to medium
Leaf:	3-angled stalk or cane, stout, topped with an umbel to 24 inches wide of super-fine, filiform leaves.
Flower:	Inconspicuous
Fruit:	Unimportant
Geographic Location:	Egypt, Sudan, etc. Not cold-hardy.
Dormant:	Evergreen
Culture:	Bog or wetland (not saline) plants, rich soil, sun or shade. Roots can be submerged in water. When dividing large clumps use smaller outside cane-clumps for planting and for elimination of larger center canes.
Use:	A natural positioning at bridge heads, also flanking waterfalls and rapids and pond margins. This plant is an unusually fine subject for its striking form, silhouette and shadow pattern.
Other:	C. haspan viviparus: a miniature, dwarf papyrus.

Dizygotheca elegantissima

THREAD-LEAF
  FALSE ARALIA

Dizygotheca: double recep-
    tacle for the
    4-celled anthers

elegantissima: top-most or
    ultimate in elegance

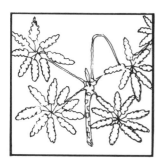

Size:	To 10-12 feet x 6-7 feet spread. Small tree.
Form:	Adult: narrowly upright like an aralia. Usually 3-4 main stems, woody below and somewhat branched.
Texture:	Juvenile is fine, adult is medium to coarse with an open character.
Leaf:	Adult: digitate, coarsely notched leaflets to 12 inches long and 3 inches wide, dark and shiny green.
Flower:	Occasionally
Fruit:	Rare
Geographic Location:	New Hebrides. Tropical Florida to 33°F.
Dormant:	Evergreen
Culture:	Give ample light, but no direct sun. Needs fast draining, moisture retentive soil. Feed 5 to 6 times per year. Does best when protected from winds. Pests: scales and nematodes.
Use:	The mature plant is a naturally japanesque subject casting a lacy shadow on a white wall or it creates an ephemeral silhouette through a glass wall surface. Subject to freezes with some damage. Recovery slow.

Dracaena arborea

DRACAENA

Dracaena: Greek for female dragon

arborea: tree-like

Size:	To 15 to 20 feet, however more like 8 x 4 feet.
Form:	Carried on a single or forked trunk 4-6 inches in diameter. A rosette or cluster of straplike leaves in a crown of 3-3½ feet in diameter.
Texture:	Medium
Leaf:	A medium-broad lanceolate leaf, 12-15 inches long x 2-2½ inches wide, dark green above.
Flower:	A large spray of flowers upright but small individually.
Fruit:	Fall berries, red, viable usually.
Geographic Location:	Brazil. Tropical Florida, to 32°F.
Dormant:	Evergreen
Culture:	Prefers a rich loam, rather moist but well-drained. Fertilize once a month in Summer. Dappled shade or full sun. Propagate by seed, cuttings or air-layering.
Use:	Both D. arborea and D. draco are interused in South Florida for the tropical effect. D. draco is most resistant to wind and salt and is a superior silhouette plant. Both are fine accent plants for mixed plant groupings.
Other:	D. draco: Dragon Dracaena; blue-green, strap-leaf, leathery, grows into a large tree.

**Dracaena deremensis
'Janet Craig'**

DRACAENA JANET CRAIG

Dracaena: Greek for
female dragon

deremensis: on a regular
monthly interval

Size:	To 10 x 7 feet spread (multi stems).
Form:	Upright, broadly multiple stems. Foliage retained from the ground up. Dark green, shiny.
Texture:	Medium
Leaf:	10-12 inches x 2 inches, slightly ribbed, dark green. Projects out at right angles to the stems, heavily foliated. Long-lasting.
Flower:	Similar to D. fragrans.
Fruit:	Rare
Geographic Location:	Garden hybrid. Tropic Florida to 32°F.
Dormant:	Evergreen
Culture:	Tolerates soil conditions, needs moisture and 3 applications of fertilizer. Adapts well to deep shade, but also will do well in sun. Rather rapid grower always presenting a good side.
Use:	More delicate in growth habit, texture and scale than D. fragrans massangeana. It forms a heavy, but graceful, 8-foot clump that is preferred as an accent in the shrub bed. Check availability re: sizes and number of canes in a clump before you plant or specify.

Dracaena fragrans massangeana

CORN PLANT

Dracaena: Greek for female dragon

fragrans: for perfume of the flowers

massangeana: a man's name

Size:	To 20 feet but more likely 4 to 10 feet x 5 feet. Usually grown from old canes where 3 are planted in one container and grown on.
Form:	Narrowly upright
Texture:	Coarse
Leaf:	Leaves are cornlike, 3 feet long x 4 inches wide, wavy marginal and variegated, in the case of massangeana, by a broadly yellow stripe in center of the leaf.
Flower:	Pinkish white, buttons issuing from a pendant raceme 2½-3 feet long, fragrant.
Fruit:	Globose berries with 1 to 3 seeds. Somewhat rare.
Geographic Location:	North East Africa. Tropical South Florida only.
Dormant:	Evergreen
Culture:	Fertile, moisture-retaining soils are essential for maintaining good foliage. Positive drainage, sunny exposure, wind protection and fertilize 3 times per year. Winter tip burn. Pests: leaf spot can be serious.
Use:	Somewhat difficult to use in natural groupings but forms an ideal wind breaker or visual screen when closely planted in line. Ideal subject for interior planters and is popular as a tubbed specimen in patios and on decks, protected.

Dracaena marginata

DRACAENA

Dracaena: Greek for
female dragon

marginata: for pink
margin pencilling

Size:	To 20 feet x 18 feet. Usually to 8-10 feet x 5-6 feet.
Form:	A woody trunk with 2-3 stems, a few branches all tipped with a ± 2-foot spread rosette of ribbon-like leaves.
Texture:	Fine
Leaf:	Long, narrow, 2 feet x ½ inch wide, pink margin.
Flower:	2 inches long in clusters, reclined at tips of branches.
Fruit:	Globose, golden.
Geographic Location:	Hawaii. Leaves damaged below 32°F, but stems survive and revive.
Dormant:	Evergreen
Culture:	Wide soil tolerance, but good drainage. Sunny or high shade. Fairly drought-resistant. Slow and durable. Cut back tips when too tall to reroot them. Tips will sprout new crown.
Use:	Selected "marginatas" carrying many canes are excellent subjects for symmetrical, espalier features. For the japanesque "character" type specimen it may be unequaled. Again, for the purpose of an accent plant both in shape as well as texture it is a joy to work with.

Dracaena reflexa

DRACAENA

Dracaena: Greek for female dragon

reflexa: leaves bent abruptly backward to 90° or more

Size:	12 feet x 8 feet.
Form:	Upright with strongly vertical main stems; lateral branches are limber and gracefully irregular.
Texture:	Medium
Leaf:	Rosette of densely clustering, short and narrow leathery leaves, deep glossy green, without midrib, wavy and reflexed, persistently clasping the willowy self-branching stem.
Flower:	White flower ¾" long sometimes reflexed. Simple or branched, short tube, long lobes.
Fruit:	Rare in Florida.
Geographic Location:	Madagascar, Mauritius, India. Tropical South Florida.
Dormant:	Evergreen
Culture:	Easily propagated from cuttings, air-layerings. Sunny to light shade, rich, moist soil. Tropical temperatures only. Fertilize 3 times per year. Positive drainage.
Use:	For a soft but sculptured accent in a shady shrub grouping it knows no peer. As a container or tubbed plant it wins applause. Use it as an espaliered subject, especially on white stucco walls, with great effect.
Other:	D. Song of India: green and white variegation.

Hymenocallis latifolia

SPIDER LILY

Hymenocallis: beautiful
     membrane as in
     the flower

latifolia: broad-leafed

Size:	2½ x 5 feet
Form:	Symmetrical outline of crown formed by strap leaves.
Texture:	Coarse
Leaf:	Dark green, lily-like strap leaf, 3 feet in length.
Flower:	Numerous, white, clustered. Slender tube to 6 inches long with narrow, long, recurving sepals and petals. Filaments upright and connected by a gossamer web.
Fruit:	Capsule, ovoid, 2 inches long.
Geographic Location:	Native to Florida coast from Melbourne to the Keys.
Dormant:	Evergreen
Culture:	Sandy loam to pure beach sand. Full sun. Average to light moisture. Usually propagated by bulb divisions. Long summer inflorescence. No serious pests.
Use:	Useful for salt-exposed locations as an accent blended into ground covers. such as Beach Sunflower, Wedelia or Beach Morning Glory.

Moraea iridioides

AFRICAN IRIS

Moraea: Latin for J. Moraeus, father-in-law of Linnaeus

iridioides: Iris-like plant

Size:	2 feet x 2 feet.
Form:	Clumps of narrow, stiff iris-like leaves that grow fan-shaped.
Texture:	Medium fine.
Leaf:	20-24 inches long, narrow, basal as in flags.
Flower:	White with yellow bands, 1½ to 2 inches long, called "Lemon Drop", the crests of the style marked with blue. Blooms every 10 days throughout the year.
Fruit:	Seed pods. Remove them to encourage more flowers.
Geographic Location:	South Africa. Central and South Florida, 25°F or higher.
Dormant:	Evergreen
Culture:	Most any fertile soil. Prefers moisture to bloom well, but is tolerant of dry soil conditions. Full sun. Fertilize three times per growing season. Lift whole plants and divide every three years.
Use:	Remember the old adage: "Combine plants in the same family". Use Moraea with a generous cover of Liriope or Dwarf Day Lilies. Try Moraea also with the Prostrate Junipers or Dwarf Ilex vomitoria. We have used it sparingly in French drains composed of white egg rock.
Other:	M. iridioides var. bicolor is called Orange Drop.

Musa acuminata 'Cavendish'

CAVENDISH OR
     CHINESE BANANA

Musa: for A. Musa, physician
     to the first
     Roman emporor

acuminata: tapering
     gradually to the apex

Size:	6-8 feet x 8-foot spread.
Form:	Soft, thickish stem or stems crowned by 6 or 7 stunning leaves arranged from 12:00 to 5:00.
Texture:	Coarse
Leaf:	5 feet long x 2 feet wide with a strong midrib. Leaf fabric is dark green and subject to tearing in a wind.
Flower:	Flower head is an oblong ball of dark red, fleshy bracts, under each bract is a cluster of inconspicuous flowers.
Fruit:	Each cluster of flowers there develops a "hand" of delicious small bananas, 6 inches long.
Geographic Location:	South East Asia. Tropic South Florida. 32°F or higher; roots will regenerate new growth if frozen to the ground.
Dormant:	Evergreen. Fast growing.
Culture:	Extremely rich loamy soil with copious water, yet well drained. It is almost impossible to fertilize bananas too heavily. Protect from winds. Remove old stalk after bearing and producing suckers.
Use:	To produce a design for a Cuban sugar planter "use bananas at the outhouse, never near the house". There should always be a space made in building a swimming pool covered with screening fabric to include a banana. Plant 2 or 3 young banana plants about 6 to 8 feet apart so that when one dies there are several others filling the space.

Neomarica coerulea

MARICA IRIS

Neomarica: new Marica

coerulia: sky blue

Size:	Mature clumps 3-3½ feet x 4½-foot spread containing 12 to 14 fans.
Form:	Fan-shaped more or less in opposite directions (such as North-South) with many Iris-like leaves.
Texture:	Medium
Leaf:	18 inches x 2½ inches tapered strap-leaf rises out of tight cluster at root line. Develops into several fan-clusters around original plant.
Flower:	2 inches in diameter, rises from the upper portion of the leaf. Fragrant. Short-lived iris-like outer petals bright sky blue; center petals ultramarine with yellow and brown pencilled markings.
Fruit:	Capsule. Rare.
Geographic Location:	South Brazil. Tropic South Florida.
Dormant:	Evergreen. Cold-hardiness not well-known, but tolerates 30-32°F.
Culture:	Of easy culture, doing well in light shade or sun. Its native eco-site is in the foothills above Santos in open fields. Tolerates some dryness and windy situations. Flowers regularly but for short periods.
Use:	It is successfully used in a bed of Liriope Big Blue in front of a colony of colored-leaf Ti plants. Also appropriate as an edge-of-pond accent-bed. Blends well with other perennial lilies.

Pennisetum setaceum

FOUNTAIN GRASS

Pennisetum: feather-like awn

setaceum: bearded or
  bristle-like

Size:	3 feet x 2½ to 3-foot spread.
Form:	The common name says it all. A graceful clump with nodding flower (bearded) stalks with a purple cast.
Texture:	Fine
Leaf:	1-2 feet long, narrow, rough curving grass blades green or purplish.
Flower:	Rose or purple, feathery, cylindrical fuzzy flowers, 6 inches long x ½ inch spikes on 3-foot stems, hollow centers.
Fruit:	Small seeds that will volunteer, forming small tuft-like many-blended seedlings.
Geographic Location:	Africa, Afghanistan. An escape on the Island of Hawaii where it volunteers profusely on old sterile lava beds. State of Florida.
Dormant:	Dormant in the Winter for a limited period; but old brown clumps, not unattractive, persist till Spring.
Culture:	Any soil, full sun; stands dry conditions, good drainage. Can be an escape under certain conditions. Shear back in the Fall.
Use:	The mini-counterpart of Pampas Grass, this subject is equally as valuable as an accent plant; its use is better than Pampas Grass when applied to smaller scaled plants and spaces. This grass often is used in plantings around fountains for its similar silhouette.
Other:	P. s. var. cupreum. Distinguished by coppery red foliage. Not as symmetrical and looser.

Philodendron giganteum

GIANT LEAF
PHILODENDRON

Philodendron: Latin for
tree-loving

giganteum: giant or very large

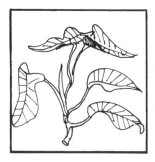

Size:	A slow climber to 6-10 feet x equal spread.
Form:	A giant aroid with succulent climbing aerial roots.
Texture:	Very coarse.
Leaf:	Beautiful lacquered leaves to 4 feet long x 20 inches wide. Cordate ovate, pale veins on closely bunched petioles.
Flower:	Typical aroid with white spathe and spadix.
Geographic Location:	Puerto Rico to Trinidad, West Indies. Coastal South Florida.
Dormant:	Evergreen. Damaged severely by 28°F, but recovers with new foliage in the Spring.
Culture:	Copious waterings are needed for gratifying results. Rich forest sandy loam. Requires dappled sun to full shade. On Trinidad this subject grows luxuriantly on large horizontal branches.
Use:	Use it at the base of rough-barked trees as in a group of Sabal Palms or in a grove of Live Oaks. Also is a dramatic tubbed specimen in lobbies or on shady terraces and patios.

Philodendron selloum

Philodendron: Greek for
    tree-loving as a climber

selloum: saddle-like

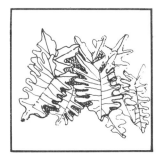

Size:	A short trunk self-header with monstera-like deeply lobed leaves, heart-shaped to 3 feet long.
Form:	Loose and open, but neatly arranged.
Texture:	Coarse
Leaf:	Stem 2-3 feet, blade 2½ feet x 2-foot spread, heart-shaped, doubly lobed and subdivided.
Flower:	White spadix semi-enclosed by thick spathe, white.
Fruit:	Fleshy, densely packed ovaries.
Geographic Location:	Central America. Tropical South Florida. Central Florida on the coasts to Sarasota and Cape Canaveral.
Dormant:	Evergreen
Culture:	Highly fertile loam with good moisture-retaining capabilities. Fertilize with high organics in intervals of 4 to 6 weeks all growing season. Prized for its sun-tolerance, but does well in dappled shade. Always mulch.
Use:	As unlikely as it may sound, P. selloum is a handsome ground cover for big scale areas, public buildings, banks, parks or condominiums. Also fitting as a subject in big planters or tubs. Not really successful as a subject for interior use.
Other:	P. speciosum, Philodendron evansii

Philodendron speciosum

Philodendron: greek for
    tree-loving as a climber

speciosum: handsome

Size:	6 to 10 feet overall x equal spread.
Form:	Arborescent, huge sagittate leaves to 6 feet long, rich green, margins wavy almost frilled.
Texture:	Coarse
Leaf:	Arrowhead-shaped to 6 feet long, thin leathery. Basal lobes notched, margins wavy.
Flower:	Fleshy spathe, green with purple margins, carmine-red inside.
Fruit:	Fleshy, densely packed ovaries.
Geographic Location:	South and Central Brazil. Tropical South Florida.
Dormant:	Evergreen
Culture:	Fertile soil, moisture-retentive, dappled shade. Organic fertilizer every 4 to 6 weeks in growing season. Protect from strong winds. Protect roots by 3 inches of mulch. Slow-growing.
Use:	The giants of Philodendrons: P. speciosum, P. eichleri and P. evansii. Use any of the above three as a powerful accent on the grand scale. For tropical effect, plant these within a large space hidden in an exotic shrub grouping; allow 3 or 4 years to show above the other plants.

Philodendron williamsi

Philodendron: Greek for
   tree-loving as a climber

williamsi: an early
   botanist, Williams

Size:	6 feet x 8 feet.
Form:	Arborescent, attenuate sagittate leaves.
Texture:	Coarse
Leaf:	Elongated, acute lanceolate, base acutely sagittate, 3½ feet long x 15 inches wide, petiole 3 feet.
Flower:	Spadix creamy white. Spathe, white.
Fruit:	Fleshy spathe, densely packed ovaries.
Geographic Location:	Bahia and Santos, Brazil. Tropical South Florida.
Dormant:	Evergreen
Culture:	Fertile soil, moisture-retentive, dappled shade, organic fertilizer every 4 to 6 weeks in the growing season. Protect roots with 3 inches of mulch.
Use:	This philodendron seems the ideal form for working with water around pools, ponds and water ways, developing a shady overstory to make the sites for planting feasible.

Polyscias fruticosa

PARSLEY-LEAF ARALIA
ALSO MING ARALIA

Polyscias: Greek for
many shadows

fruticosa: Latin for
shrubby or bushy

Size:	6-8 feet
Form:	Compact upright often with multi-upright leads. Always with strong verticality.
Texture:	Fine, almost lacy.
Leaf:	Dark green, tri-pinnately, deeply cut and lobed, narrow-toothed, to 4 inches long, ovate-lanceolate.
Flower:	Not seen in Florida.
Fruit:	Not seen in Florida.
Geographic Location:	Polynesia, India. Coastal South Florida.
Dormant:	Evergreen. Not cold-hardy.
Culture:	Best in partial to dark shade, indirect light. Of easy cultivation, but interior plants can be overwatered and over fertilized. Slow-growing and needs little shaping. Pests: watch for mites and scales.
Use:	An Aralia with a distinctly japanesque appearance to be used thus. For patio, porch, and bathroom gardens, it is a must. Lives for years as a tubbed or potted accent. A patrician.
Other:	P. fruticosa elegans: a small-growing, very compact subject.

Spartina bakeri

SAND CORDGRASS

Spartina: pertaining to the
    broom genus

bakeri: named for American
    botanist John Baker,
    1834-1920

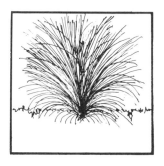

Size:	3 to 4 feet, even wider spread, rounded.
Form:	Robust bunch grass. Old plants often form bunches 18-20 feet in diameter.
Texture:	Fine
Leaf:	Rolled when drying, to ¼ inch wide; upper surface dark green, lower light green. Upper surface ridged.
Flower:	Early Spring. Obscure, relatively scarce.
Fruit:	Seed-head 2 to 8 inches long, 5 to 12 spikes, each 1¼ to 2½ inches long.
Geographic Location:	From South Florida to Carolinas. Sandy soils on margins of swales, pinelands. St. John's River Basin.
Dormant:	Some basal leaves remain green during Winter in Florida.
Culture:	Margins of sand ponds and fresh water marshes, tolerates periodic flooding during growing season. Not for saline soils. Reproduces mainly by rhizomes. Should be evergreen in South Florida.
Use:	Excellent for retention basins being tolerant to both dry and flooding conditions. A fine native for shoreline use, for ponds and streams. Combines with rock outcroppings. Fine for bridgehead.

Spathiphyllum 'Mauna Loa'

PEACE LILY

Spathiphyllum: Greek for
    spathe-like leaf

Mauna Loa: named for
    volcano in Hawaii

Size:	3 feet x 2½ feet.
Form:	Lanceolate plantain-like leaves atop slender petioles arranged in a dense cluster, all arching to the sides.
Texture:	Medium
Leaf:	Dark green 2-foot leaf, elliptical tapering to a point atop a petiole at the same length.
Flower:	A white spathe and slender spadix as long as 15 inches. The spathe turns jade green in time.
Fruit:	Fleshy small ovaries tightly packed. Berry-like.
Geographic Location:	This is a hybrid derived from parents of the American tropics. South and Central Florida.
Dormant:	Evergreen
Culture:	Rich, high fertility soil with moisture-retentive qualities. Positive drainage, high shade or dappled sun. Protect from strong winds. Fertilize every 4 to 6 weeks throughout the growing season. No direct sun.
Use:	The generous production of white flowers offers a welcome break from "tropical greenery". Use it in combination with facer plants like Liriope as the second tier ground cover. Don't overlook its value as a group accent in a cover bed of prostrate Junipers, Holly Fern or Peperomia.
Other:	S. clevelandi: dwarf, 12-14 inches x 12 inches, similar.

285

**Sphaeropteris cooperi
(alsophila australis)**

AUSTRALIAN TREE FERN

Sphaeropteris: Greek for fern
with globular sporangia.

cooperi: botanist of the
last century

Size:	Ultimate size to 20 feet with fronds to 12 feet long.
Form:	Upright-grower with feather-shaped fronds forming a handsome crown. Trunk to 12 inches in diameter, wooly, russet.
Texture:	Fine.
Leaf:	Fronds from juvenile stage from 1 foot to 12 feet spread in old specimens. Light bright green.
Flower:	None.
Fruit:	Spores on undersides of leaves.
Geographic Location:	Australia. Central Florida to South Florida.
Dormant:	Evergreen, is hardy to 25°F or slightly lower for short periods.
Culture:	Fertile sandy loam, well-drained. Dappled to full shade. Requires regular moisture with frequent overhead misting. Light but regular fertilizing during Summers. Pests: mites and mealy bugs.
Use:	With careful siting, this very tropical giant fern will present a really Hawaiian atmosphere. They are useful as subjects around a shaded pool or pond for the double image created on the water's surface.

286

Strelitzia nicolai

WHITE BIRD OF PARADISE

Strelitzia: wife of
    King George III

nicolai: after a
    Russian nobleman

Size:	To 30 feet x 15 feet.
Form:	Tree-like clumping with a display of banana-like leaves with staggered trunks.
Texture:	Coarse
Leaf:	Gray green banana-like, 5-10 feet long arranged fanlike on erect trunks.
Flower:	White with dark blue tongue, bract is red-brown, 6 inches high x 18 inches long.
Fruit:	3-angled capsules that split to release seeds.
Geographic Location:	South Africa. Tropic South Florida.
Dormant:	Evergreen. To 28°F with some leaf damage, but will recover.
Culture:	Full sun to high shade. Good drainage. Feed young plants to push maturity at which point terminate feeding to limit need for pruning. Water regularly, protect from winds. Subject to scales.
Use:	This is a very special accent plant of big scale. Use it for an entrance guardian. A perfect plant for the swimming pool where it won't be messy. One of the hardiest accent plants for large scale projects in Florida.

287

Strelitzia regina

BIRD OF PARADISE

Strelitzia: for the wife of
King George III

regina: Latin for queen

Size:	4 feet.
Form:	A cluster of waxen paddle-shaped leaves with a boat-shaped large inflorescence.
Texture:	Coarse
Leaf:	About 8 leaves per plant. Their blades narrow ovate, leathery, less than 18 x 6 inches exceeding their stems.
Flower:	Stands as high as the leaves in a bract which is green, orange sepals and blue tongue, 3 inches long rise above the bract.
Fruit:	3-angled capsules that split to release seeds which are red.
Geographic Location:	Cape of Good Hope. South and Central Florida. 28°F.
Dormant:	Evergreen
Culture:	Sunny to high shade. Rich, moisture-retentive soil. Water faithfully. Good drainage. Fertilize every 4 to 6 weeks in the growing season. Pests: scales and grasshoppers.
Use:	Accent plant extraordinary: for form, for foliage and texture and finally for inflorescence. This plant will blend with a great many ground covers when used as a landscape focus.

Tripsacum dactyloides

FAKAHATCHEE GRASS

Tripsacum: triple sweet

floridanum: of Florida,
    the State

Size:	To 4 feet with a spread to 6 feet.
Form:	Rigid leaf blades forming a radial hemisphere in profile.
Texture:	Fine.
Leaf:	To 3 feet. Spread of leaf ¾ inch, tapers at tip as well as at the base, flat blade.
Flower:	Staminate portion of spike of flowers displays rust-colored anthers, 4 inches long, Spring and Summer.
Fruit:	Female (lower portion) of spike develops in to bony joints (grains) as in a finger, 2 inches long.
Geographic Location:	Southern peninsula of Florida, in the Keys as well as the Everglades and Big Cypress Swamp. Low rocky pinelands in shallow soil.
Dormant:	Grassy parts subject to freezes. Comes back in Spring.
Culture:	This true native of Southern Florida adapts to dense and light soils and is tolerant of wet soil conditions. Sunny sites only. No known pests.
Use:	This plant, along with Fountain Grass and Pamapas Grass, etc. form a refreshing accent subject for ground cover beds as well as stream and pond margins. Edges of hammocks a typical environment.
Other:	T. dactyloides, Gama Grass: 9-foot high image of T. floridanum.

Yucca elephantipes

SPINELESS YUCCA

Yucca: modification of an aboriginal name

elephantipes: refers to root-swelling

Size:	To 30 feet. More like 8 to 12 feet locally.
Form:	Clump or single trunk, low branching form a pleasing arrangement with a swelling at ground line.
Texture:	Coarse
Leaf:	Dagger-like but soft-tipped to 4 feet long x 3 inches wide, dark green, rough leaf margins.
Flower:	White petals hanging from erect panicles, candelabra-like. Edible in salads or breaded and fried.
Fruit:	Deep brown seed capsules.
Geographic Location:	Mexico, Central America. Central and South Florida. 26-28°F.
Dormant:	Evergreen
Culture:	Any soil condition, but with plenty of moisture with fertilizer forms a stunning scultured shape. Positive drainage. Full sun or high shade. Protect from winds. Subject to leaf-staining fungus.
Use:	This subject ranks high in the number of times it is used in Florida as an accent comparatively speaking. Against stark concrete walls it is superb. As silhouetted against a fixed glass you'll be surprised.
Other:	Y. aloifolia: native to Florida dunes, called Spanish Bayonette, sharp-pointed leaves, candelabra of white flowers.

Zamia furfuracea

SCURFY ZAMIA

Zamia: Latin for Pine nut

furfuracea: bran-like or
    scurfy particles

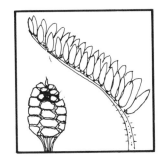

Size:	3½-4 feet x 7 feet.
Form:	Loose spreading symmetrical rosette.
Texture:	Coarse
Leaf:	Leaflet obovate, 5 inches long x 1½ inches wide. Leaf 3-4 feet x 9 inches wide, dark green, woody, veins many parallel, longitudinal.
Flower:	Cone male and female, dioecious.
Fruit:	Aggregate cone. Tightly packed, fleshy, crimson seeds exposed, 3/4 inch x 3/8 inch diameter.
Geographic Location:	Mexico. South Florida, to 25°F.
Dormant:	Evergreen
Culture:	Sandy soils. Prefers full sun, well-drained condition. Tolerant to drought after establishment. Good salt-resistance. Medium growth. Pests: scale insects.
Use:	Makes a strong statement as a ground cover, but it is too often spaced closely, not allowing for ample spread. As an accent plant it ranks high for use in large scale, sunny areas where either ground covers or mixed coarse-textured plantings are used.

INDEX

BIBLIOGRAPHY

AUTHOR	TITLE	PUBLISHER	CITY	YEAR
Alcoa Steamship Co., Inc.	Flowering Trees of the Caribbean	Rhinehart and Co.	New York	1955
Alfred B. Graff	Exotica Series 3	Roers & Co., Inc.	E. Rutherford, N.J.	1973
Bell & Taylor	Florida Wild Flowers	Laurel Hill Press	Chapel Hill, N.C.	1982
David Sturrock & E. A. Menninger	Shade & Ornamental Trees of The Caribbean	Stuart Daily News	Stuart, Florida	1946
Editors of Sunset Books & Sunset Magazine	Sunset New Western Garden Book	Lane Publishing Co.	Menlo Park, CA	1967
Erdman West & Lillian E. Arnold	The Native Trees of Florida	University of Florida Press	Gainesville, Florida	1952
Flemming, Genelle & Robert Long	Wild Flowers of Florida	Banyan Books, Inc.	Miami, Florida	1976
Florida Division of Forestry, Florida Department of Agriculture	Coastal Plants of Florida	Florida Department of Agricultural & Consumer Services	Tallahassee, Florida	1979
Florida Division of Forestry, Florida Department of Agriculture	Forest Trees of Florida	Florida Department of Agriculture & Consumer Services	Tallahassee, Florida	1972
Florida Division of Forestry, Florida Department of Agriculture	Urban Trees of Florida	Florida Department of Agriculture & Consumer Services	Tallahassee, Florida	1979
Florida Department of Natural Resources	Aquatic & Wetland Plants of Florida	Florida Department of Natural Resources	Tallahassee, Florida	1979
George Stevenson	Palms of Florida	George Stevenson	Miami, Florida	1974
George Stevenson	Trees of The Everglades National Park & Florida Keys	George Stevenson	Miami, Florida	1969
Marie C. Neal	In Hawaiian Gardens	Bishop Museum	Honolulu, Hawaii	1948
Mary Francis Baker	Florida Wild Flowers	The Macmillan Co.	New York	1949
P.S. Tomlinson	Biology of Trees Native To Tropical Florida	Harvard University Printing Office	Allston, Mass.	1980
Staff, L.H. Bailey Hortotium	Hortus Third	Macmillan Company	New York	1976
Small, J.K.	Manual of the Southeastern Flora	Habner Publishing Co.	New York	1972
Texas Forest Service	Forest Trees of Texas	Texas Forest Service Texas A&M College System	College Station, TX	1958
U.S. Department of Agriculture	Plants for the Coastal Dunes	U.S. Department of Agriculture, Soil Conservation Service	Washington, D.C.	1984
Watkins and Sheehan	Florida Landscape Plants	University Presses of Florida	Gainesville, Florida	1975

Williams, R.O.	Useful & Ornamental Plants of Trinidad & Tobago	Guardian Commercial	Port of Spain, Trinidad	1951
Workman, Richard	Growing Native	The Sannibel-Captiva Conservation Foundation	Sanibel, Florida	1980
Wyman, Donald	Ground Cover Plants	Macmillan Company	New York	1956
Georgia Tasker	Wild Things	Florida Native Plant Society	Winter Park, Florida	1984
Harrar & Harrar, E.S. & J.G.	Guide to Southern Trees	Dover Publications, Inc.	New York	1962
H.F. Macmillan	Tropical Planting & Gardening	Macmillan & Co., Ltd.	New York	1956
Horace Clay & J. Hubbard	Trees for Hawaiian Gardens	University of Hawaii Extension Service	Honolulu, Hawaii	1962
Julia Morton	500 Plants of South Florida	E.A. Seemann Publishing Company	Miami, Florida	1974
Kuck and Tongg	The Modern Tropical Garden	Tongg Publishing Co.	Honolulu, Hawaii	1955
Long and Lakela	A Flora of Tropical Florida	University of Miami Press	Miami, Florida	1971